MOLECULAR EVOLUTION

MOLECULAR EVOLUTION

EDITED BY FRANCISCO J. AYALA
UNIVERSITY OF CALIFORNIA, DAVIS

To Mitzi

The Cover

Starch gel showing variation in the enzyme *acid phosphatase* in flies of the tropical species *Drosophila equinoxialis*. Flies homozygous at the gene locus coding for the enzyme exhibit one band; heterozygous flies show three bands.

MOLECULAR EVOLUTION

First printing

Sunderland, Massachusetts 01375

Manufactured in the U. S. A.

Library of Congress Catalog Card Number: 75-36113

ISBN: 0-87893-044-2

CONTENTS

PREFACE

The spectacular achievements of molecular biology during the last two decades have towered over those of any other biological field, and indeed of any other science. The scientific community has been astounded again and again by dramatic discoveries concerning the molecular processes and organization of living beings. These accomplishments have borne multifarious and often unforeseen fruits both in other branches of science and in applied fields such as human health. The consequences for evolutionary biology have been most impressive.

The systematic application of the conceptual models and techniques of molecular biology has provided new insights into evolutionary processes, and new methods in the reconstruction of evolutionary history. Techniques such as gel electrophoresis, immunodiffusion, protein sequencing, and DNA hybridization have become powerful means to study biological evolution. Indeed, the last 10 years have witnessed the blossoming of a new field of study whose subject is biological evolution at the macromolecular level. This new biological discipline might be called "Molecular Evolution."

It seems appropriate at this time to attempt a summary and synthesis of the evolutionary knowledge obtained by molecular studies. In the field of molecular evolution an era—which could be named the "Age of Consolidation"—has ended, and a new era—the "Age of Expansion"—has started. Some basic parameters (such as the degree of genetic polymorphism in natural populations and the amount of genetic change during speciation) and various methodologies (such as amino acid sequencing to determine phylogenetic history) became established during the Age of Consolidation. The time is ripe now to review these accomplishments. This could provide stimulus and guidance to future molecular evolutionary studies. The dawning Age of Expansion should not just accumulate more and more data to define more precisely parameters already established, nor should it only add further sophistication to methods already validated. Expansion will be most valuable if it is in depth and advances the field conceptually.

There are additional reasons why a summary of the accomplishments of molecular evolution is appropriate. Studies of molecular evolution are published in journals with subject matter as diverse as evolution, genetics, systematics, molecular biology, and biochemistry. Few scientists and even fewer students are exposed to such a gamut of journals. Molecular biology uses distinctive terminology and methodologies. Some biologists have not followed the advances of molecular evolution because they felt unable to understand them. Such mystification is unwarranted. Only moderate effort is required to understand the basic concepts, nomenclature, and methods used in molecular evolution.

This book is intended to satisfy all these needs. I organized a symposium entitled "Molecular Study of Biological Evolution," held at the University of California, Davis, on June 17 and 18, 1975. The symposium was sponsored by the Society for the Study of Evolution and the American Society of Naturalists. It was attended by some 500 scientists and students from the United States and abroad. The invited speakers were asked to write their contributions, not in the style of research papers or review articles, but rather as chapters for a book intended for biologists of all persuasions, as well as for biology students and for the "intelligent public" at large.

The broad spectrum of materials covered in this volume are not presently available anywhere else under a single cover. This book should, then, be useful to a very diverse public. Scientists will find here a summary of the field of molecular evolution. The book can also be read by nonscientists interested in the spectacular new advances made in the theory of evolution through molecular studies. The volume should be particularly useful to biology students. Textbooks of evolution lack adequate presentation of recent molecular advances. The book can be used as a supplementary text for undergraduate as well as graduate courses in evolution, genetics, molecular biology, biochemistry, and even general biology. It will provide the ideal text for courses and seminars specifically dedicated to molecular evolution.

The advantages and disadvantages of multiple authorship weigh unequally in different circumstances. In the present case, the main advantage can be simply stated: No single person could have treated authoritatively all the subjects covered in this book. A book with many authors has such disadvantages as unevenness of style and level, repetition, gaps, inconsistencies. I have used liberally the privileges and duties of an editor to remedy these problems. I have unified the format, made deletions, and on occasion rewritten parts of the original contributions. I have not, however, changed the opinions of the authors even when they contradicted one another. Science is a dialectical enterprise. Alternative hypotheses are desirable, and even necessary, in an advancing field. The reader will notice particularly that the authors of Chapters 2 to 5 favor different hypotheses about the factors maintaining protein polymorphisms in natural populations.

The first chapter of the book is a brief introduction to molecular genetics and to some basic evolutionary concepts. It also includes brief reviews of some molecular evolutionary topics, such as gene duplications, that did not call for a full chapter in this book. Chapters 2 to 5 deal with molecular variation *in* populations. These chapters belong together for reason of their subject matter and also because of the predominant experimental methodology involved—gel electrophoresis. The rest of the book is concerned with variation *between* populations. Chapter 6 introduces the concept of species and compares the accomplishments of premolecular and molecular studies of species formation. The genetic changes concomitant to speciation are discussed in Chapter 7 for animals and in Chapter 8 for plants. Gel electrophoresis is again the methodology of the studies reviewed in these two chapters. Chapters 9 and 10 consider the application of techniques such as amino acid sequencing of proteins and immunodiffusion to the reconstruction of phylogenetic history and to related problems such as taxonomic relationships, rates of evolution, and the like. Chapter 11 reviews the DNA content of organisms and draws inferences about evolutionary processes in the geological time scale. The organization of DNA, particularly with respect to unique and repetitive DNA sequences, is the subject of Chapter 12. Chapter 13 contrasts the evolution of structural genes with the evolution of gene regulation; the latter may, in fact, play the critical role in the evolution of morphology, function, and reproductive affinities.

I am greatly indebted to the authors for the considerable effort and expertise invested in the preparation of their chapters and for following the guidelines received from the editor. They were extremly tolerant of the extensive editorial changes that I made in some places. The individual chapters are excellent. It is my hope that the book will be useful to students of evolution and to scholars, and that most readers will agree that in this case the whole is more than the sum of its parts.

FRANCISCO J. AYALA
Davis, California
January 15, 1976

CONTRIBUTORS

JOHN C. AVISE, Department of Zoology, University of Georgia, Athens

FRANCISCO J. AYALA, Department of Genetics, Unversity of California, Davis

ROY J. BRITTEN, Division of Biology, California Institute of Technology, Pasadena, and Staff Member, Carnegie Institution of Washington

MARGARET E. CHAMBERLIN, Division of Biology, California Institute of Technology, Pasadena

ERIC H. DAVIDSON, Division of Biology, California Institute of Technology, Pasadena

THEODOSIUS DOBZHANSKY, Department of Genetics, University of California, Davis

WALTER M. FITCH, Department of Physiological Chemistry, University of Wisconsin Medical School, Madison

GLENN A. GALAU, Division of Biology, California Institute of Technology, Pasadena

MORRIS GOODMAN, Department of Anatomy, Wayne State University School of Medicine, Detroit

LESLIE D. GOTTLIEB, Department of Genetics, University of California, Davis

RALPH HINEGARDNER, Division of Natural Sciences, University of California, Santa Cruz

BARBARA R. HOUGH, Division of Biology, California Institute of Technology, Pasadena

GEORGE B. JOHNSON, Department of Biology, Washington University, St. Louis

ROBERT K. SELANDER, Department of Biology, University of Rochester, Rochester

MICHAEL SOULÉ, Department of Biology, University of California, San Diego

JAMES W. VALENTINE, Department of Geology, University of California, Davis

ALLAN C. WILSON, Department of Biochemistry, University of California, Berkeley

CHAPTER 1

MOLECULAR GENETICS AND EVOLUTION

FRANCISCO J. AYALA

DNA—THE GENETIC MATERIAL

The process of biological evolution consists of changes in the genetic constitutions of organisms. Each individual experiences throughout its life multiple changes in morphology, physiology, behavior, and the like. These changes lack permanence; they disappear with the individual. Changes in the hereditary materials, however, may be passed from one individual to its descendants and thus are cumulative over the generations. The stupendous evolutionary changes occurring throughout the history of life, as well as the remarkable diversity of living organisms, are the result of accumulated genetic changes.

The genetic information is encoded for most organisms in the chemical substance known as deoxyribonucleic acid (DNA); in some viruses it is encoded in a related substance called ribonucleic acid (RNA). DNA was first extracted and described by F. Miescher around 1870. The identification of DNA as the carrier of genetic information was a gradual process resulting from several critical studies. In 1944, Avery, MacLeod, and McCarty obtained highly purified DNA from a virulent strain of a bacterium, *Diplococcus pneumoniae*. Nonvirulent bacteria were incubated in the presence of this purified DNA, and some virulent bacteria were recovered. The change from nonvirulence to virulence was shown to be hereditary; DNA was described as a *transforming* agent. Mirsky and Ris (1949) found that all somatic cells of a given organism contain, as a rule, the same amount of DNA, while gametic cells contain half as much DNA as somatic cells. This is what would be expected of the hereditary material. Direct evidence of DNA as the genetic material was obtained by Hershey and Chase in 1952. They demonstrated that when the virus *T2* infects the bacterium *Escherichia coli,* only the DNA of the virus enters the bacterium and brings about its own replication. The protein coat which envelops the virus DNA does not enter the bacterium or participate in the replication process.

DNA and RNA are long chains (polymers) composed of four kinds of nucleotides. Each nucleotide consists of a nitrogen-containing base linked with a five-carbon sugar (pentose) and a phosphate group. The pentose sugar is deoxyribose in DNA and ribose in RNA. Each phosphate group is covalently bound to the 5′ carbon of the sugar and establishes an additional covalent bond with the 3′ carbon of the sugar of a second nucleotide. The polynucleotide is held together by these 5′–3′ covalent ester bonds. The nitrogen bases in the nucleotides are of two classes, purines and pyrimidines. The two most commonly occurring purines in either DNA or RNA are adenine (A) and guanine (G). The two most common pyrimidines in DNA are cytosine (C) and thymine (T); RNA contains uracil (U) instead of thymine.

Watson and Crick proposed in 1953 that DNA exists as a double-helical molecule made up of two complementary polynucleotide chains, with the phosphate-sugar backbone on the outside and the nitrogen bases inside (Figure 1). The complementary chains are held together by hydrogen bonds between bases in different chains in such a way that A always pairs with T and G always pairs with C. It follows that DNA has as many A as T bases and as many G as C bases; that is, A/T and G/C are each equal to 1. In contrast, $(A + T)/(G + C)$ may vary from organism to organism. These properties of ratios between bases had already been shown by Chargaff (1951). The double-helix model of DNA provided a plausible explanation of the basic properties of the hereditary material—binary replication and carrier of genetic information. The double-helix model of DNA proposed by Watson and Crick has been corroborated down to almost its finest details.

The genetic information is encoded in the sequence of the nitrogen bases in the double polynucleotide chain. The four bases may be considered as the letters of the genetic alphabet. Specific sequences of letters of the English alphabet make up words; a sequence of words conveys information. In an analogous fashion, one may think of nucleotide sequences as genetic "words." The genetic endowment of an individual may then be considered to be like a "book" made up of DNA words. There is practically no limit to the number of different words or information messages that can be encoded in long DNA chains.

The basic units of information are not individual bases, but non-overlapping groups of three consecutive bases. Because there are four bases, the number of possible different groups of three bases is $4^3 = 64$. Because of the redundancy of the genetic code (see below), there are only 21 different units of information among the triplets. A polynucleotide

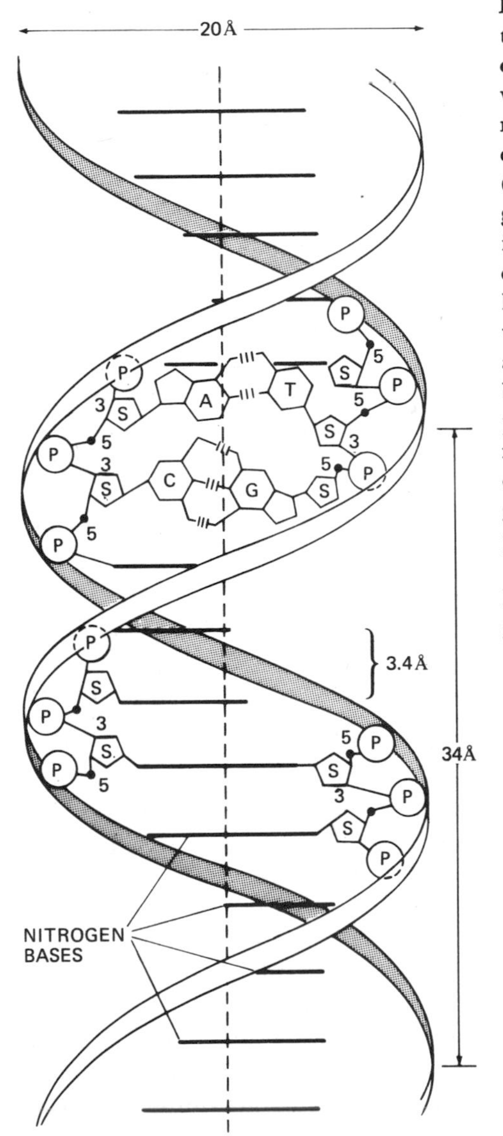

FIGURE 1. Schematic representation of the double-stranded helical configuration of DNA. The outward backbone of the DNA molecule consists of two chains each made of deoxyribose sugar (S) alternating with prosphate groups (P). Each phosphate group is covalently bound to the 5′ carbon of the sugar and establishes an additional covalent bond with the 3′ carbon of the sugar of a second nucleotide. Nitrogen bases connected to the sugars project toward the center of the molecule. The two nucleotide chains are held together by hydrogen bonds between complementary purine-pyrimidine bases. Adenine (A) and thymine (T) form two hydrogen bonds; cytosine (C) and guanine (G) form three. The base pairs are stacked flat one above the other at intervals of 3.4 Å (1 angstrom, Å, = 10^{-7} millimeter) and rotated 36°. Each chain makes a complete rotation every 34 Å, and there are 10 base pairs per complete rotation. In cross section, the two chains are separated by 120° in one direction and 240° in the other direction around the circumference of an imaginary cylinder containing the double-stranded helix. As a consequence there are two grooves along the sides of the molecule, one wider than the other. The two complementary chains run in opposite directions.

chain with 600 nucleotides (a probable typical length for the informational segment of a structural gene) has 200 nonoverlapping groups of three bases. The number of potentially different messages contained in chains of that length is $21^{200} = 10^{264}$, a number immensely greater than the number of atoms in the known universe.

The complementarity of the two polynucleotide chains in the double helix provides the specificity for the precise replication of genes. The sequence of bases along one of the strands unambiguously specifies the sequence along the complementary strand because of the strict determination of the base pairing (C with G, A with T) between the two DNA chains. Replication starts with the gradual separation of the two complementary chains by an unwinding process. Each chain then serves as a template for a complementary strand. Two new double helices are formed which because of the rules of base pairing are identical to each other and to the parental double helix.

Transcription and translation

The genetic information contained in the DNA directs the development and metabolism of the organism through the processes of *transcription* and *translation* (Figure 2). Two kinds of genes may be distinguished: Structural genes are those whose RNA products are translated into proteins (polypeptides). All other genes are regulatory; these include the genes coding for ribosomal RNA (rRNA) and transfer RNA (tRNA), as well as those whose RNA products are involved in the initiation, regulation, or termination of the transcription of other genes. (Structural and regulatory genes are not defined in the same way by all authors; the definitions given here are useful for the purpose of the following discussion.)

The process of transcription takes place in the nucleus (except for mitochondrial and chloroplast genes) and is the same for structural and for regulatory genes. The base sequence of a DNA segment is transcribed into a complementary RNA sequence. The nitrogen bases in the DNA template specify the complementary bases in RNA, according to the same rules of pairing as in the replication of DNA except that uracil rather than thymine pairs with adenine. Only one of the two strands is transcribed, but the strand transcribed need not be the same for all genes, as has been shown for the bacteriophages *T4* and λ.

The RNA transcribed from structural genes is called messenger RNA (mRNA). The mRNAs carry the information for the specific sequence of amino acids in polypeptides. Polypeptides are chains of many amino acids linked by peptide bonds joining the carboxyl (—COOH) terminal group of one amino acid to the amino ($—NH_2$) group of the next amino acid. Although more than 200 amino acids are known, only 20 are common constituents of polypeptides. The sequence of amino acids in a polypeptide is called its *primary* structure. Polypeptides also have a helical structure called the *secondary* structure of the protein. The three-dimensional configuration of the polypeptide chain is called its *tertiary* structure. Under normal physiological conditions the amino acids of a polypeptide interact with one another through

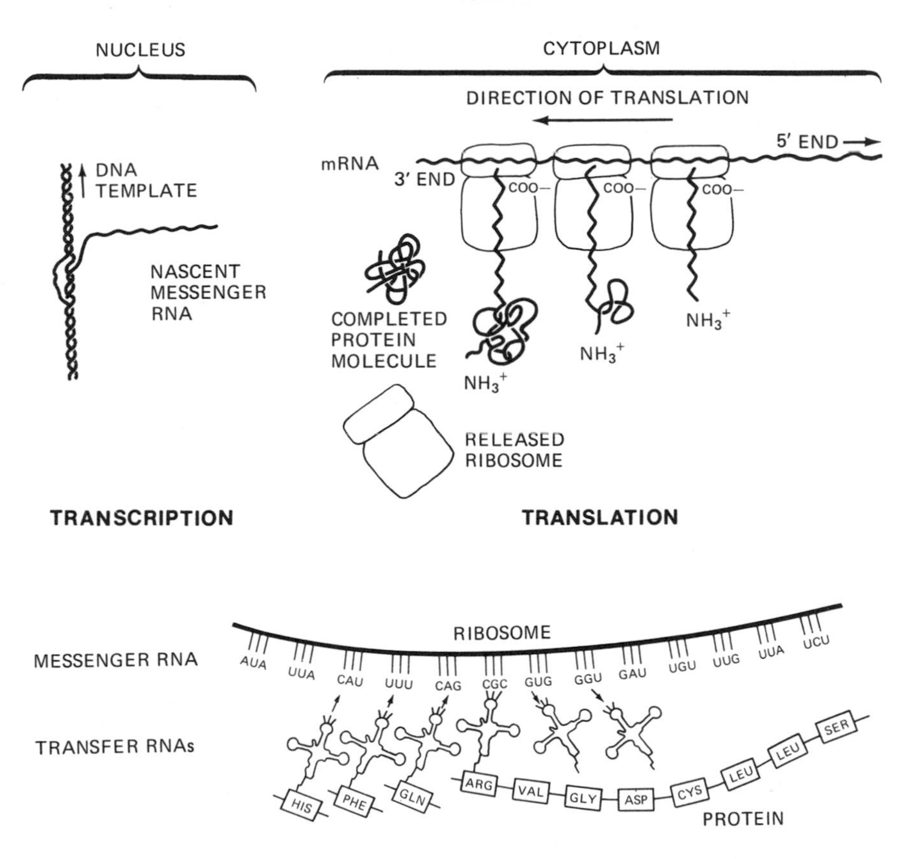

FIGURE 2. Schematic representation of the processes of transcription and translation. One strand of the DNA helix serves as a template for the synthesis of a complementary chain of messenger RNA. The messenger RNA (mRNA) moves to the cytoplasm, where several ribosomes attach to it. Each ribosome synthesizes a polypeptide, proceeding about 75 nucleotides apart from the previous ribosome. Each codon in the messenger RNA is recognized by a complementary anticodon on a transfer RNA molecule carrying a particular amino acid.

hydrogen bonds (H—H), disulfide bonds (S—S), and other kinds of bonds to produce the final biologically active configuration. The secondary and tertiary structures of a polypeptide are thus determined by its primary structure. One or more polypeptides make up a protein. Proteins consisting of more than one polypeptide have a *quaternary* structure which refers to the three-dimensional topology of the associated polypeptides. The hemoglobins are typical quaternary structure proteins. The most common form of hemoglobin found in human adults, hemoglobin A, consists of two alpha polypeptides and two beta polypeptides. One gene codes for the alpha, and another gene for the beta polypeptides.

The process by which the information encoded in the base sequence of mRNA is used to direct the synthesis of a polypeptide is called translation (Figure 2). The process of translation takes place in the cytoplasm and is mediated by ribosomes, transfer RNA molecules, and several enzymes. The mRNA synthesized in the nucleus moves to the cytoplasm, where it becomes associated with groups of ribosomes. There the base sequence in the mRNA is "read" with the aid of tRNAs that are specific for each kind of amino acid. The mRNA is translated in a stepwise fashion, starting at the 5′–carbon end and proceeding toward the 3′–carbon end. Polypeptides are synthesized proceeding from the amino end of the first amino acid to the carboxyl terminus of the last amino acid.

The information is contained in mRNA in nonoverlapping sequences of three nucleotides, called *codons.* Each codon corresponds to a complementary sequence of three nucleotides, called an *anticodon,* at a particular site of an appropriate tRNA. Each kind of tRNA molecule associates with a specific amino acid. As succeeding codons of a mRNA pass through the ribosomes, tRNAs are sequentially involved in codon-anticodon pairings. The corresponding amino acids are thus brought into position, peptide bonds form between adjacent amino acids, and a polypeptide is synthesized. There are about 60 varieties of tRNAs in higher organisms and at least 30 in the bacterium *Escherichia coli.* Because only 20 amino acids are normally involved in protein synthesis, several types of tRNA often have affinity for the same amino acid. There are, for example, five species of tRNAs in *E. coli* which have affinity only for the amino acid leucine.

THE GENETIC CODE

A single mRNA molecule is usually translated simultaneously into several polynucleotide chains by different ribosomes which work at a distance of about 75 nucleotides from each other. Each ribosome that attaches to a mRNA molecule synthesizes a polypeptide. The synthesis of a polypeptide is concluded when a ribosome encounters a *terminator* codon in the mRNA. There are three terminator codons, UAA, UGA, and UAG, none of which has affinity for any of the normally occurring tRNAs. Table I shows the correspondence between the 64 possible triplets in mRNA and the 20 amino acids. (It also lists the codons that signal for termination of protein synthesis.) The code is degenerate, because a given amino acid may be specified by more than one codon.

Different codons for a given amino acid are sometimes recognized by a single species of tRNA. Pairing takes place because some wobble is possible in the third position of a codon once the correct base pairs have formed in the first two positions. The several species of tRNA that exist for a given amino acid sometimes have an affinity for only one codon and sometimes for different codons. Thus, there is no simple one-to-one relationship between the number of tRNA species and the number of different codons in the mRNAs of an organism.

The genetic code is redundant in the direction of nucleic acid to protein, because two or more codons may code for the same amino acid. In the direction of protein to nucleic acid the genetic code is ambiguous because individual amino acids may be encoded by more than one nucleotide triplet. Knowledge of the amino acid sequence of a protein does not allow unambiguous determination of the nucleotide sequence coding for it.

TABLE I. The genetic code. Amino acids* (or termination signals) specified by each of the 64 nucleotide triplets in messenger RNA.

FIRST LETTER	SECOND LETTER: U	C	A	G	THIRD LETTER
U	UUU UUC } Phe UUA UUG } Leu	UCU UCC UCA UCG } Ser	UAU UAC } Tyr UAA Chain End UAG Chain End	UGU UGC } Cys UGA Chain End UGG Trp	U C A G
C	CUU CUC CUA CUG } Leu	CCU CCC CCA CCG } Pro	CAU CAC } His CAA CAG } Gln	CGU CGC CGA CGG } Arg	U C A G
A	AUU AUC AUA } Ile AUG Met	ACU ACC ACA ACG } Thr	AAU AAC } Asn AAA AAG } Lys	AGU AGC } Ser AGA AGG } Arg	U C A G
G	GUU GUC GUA GUG } Val	GCU GCC GCA GCG } Ala	GAU GAC } Asp GAA GAG } Glu	GGU GGC GGA GGG } Gly	U C A G

*The names of the amino acids abbreviated in the Table are as follows: Ala, alanine; Arg, arginine; Asn, asparagine; Asp, aspartic acid; Cys, cysteine; Gly, glycine; Glu, glutamic acid; Gln, glutamine; His, histidine; Ile, isoleucine; Leu, leucine; Lys, lysine; Met, methionine; Phe, phenylalanine; Pro, proline; Ser, serine; Thr, threonine; Tyr, tyrosine; Trp, tryptophan; Val, valine.

The genetic code is apparently universal, or nearly so. A given codon is always translated into the same amino acid in different organisms, with only minor exceptions known. Lane, Marbaix, and Gurdon (1971) have provided most impressive evidence showing that the genetic code is the same in different organisms. Purified mRNA coding for hemoglobin was extracted from rabbits and injected into frog oocytes; rabbit hemoglobin was synthesized there. The information in the rabbit mRNA was precisely recognized by the tRNA's of the frog oocytes, although the oocytes are normally never involved in hemoglobin synthesis. The universality of the genetic code suggests that the genetic code in its present form became fixed in some primordial form of life from which most living organisms have evolved.

THE ORIGIN OF HEREDITARY VARIATION

The process of evolution depends on the occurrence of hereditary variation. If DNA replication were always perfect, life could not have evolved and diversified; the same kinds of organisms, and no others, would be living today that existed 3 billion years ago, unless these had become extinct in the meantime.

Changes in the hereditary materials are known as *mutations.* Gene, or *point,* mutations are those that change only one or a few nucleotides in a gene. *Chromosomal* mutations are those changing the number of chromosomes or the number or arrangement of genes in chromosomes. Chromosomal mutations may be of the following kinds:

1. Those that decrease (*deletions*) or increase (*duplications*) the number of genes in the chromosomes
2. Those due to rearrangement of genes in the chromosomes: *inversions* (when a block of genes rotates 180° within a chromosome) and *translocations* (when blocks of genes change their location in the chromosomes)
3. Those that change the number of chromosomes: *fusion* (when two nonhomologous chromosomes fuse into one), *fission* (when a chromosome splits into two), *aneuploidy* (when not all the nonhomologous chromosomes occur an equal number of times), *haploidy* (when there is only one set of chromosomes), and *polyploidy* (when there are more than two sets of chromosomes)

A point mutation occurs when one nucleotide is altered and the new nucleotide sequence is passed on to the offspring. The change may be due to the substitution of one or more nucleotides for others or to the addition or deletion of one or more nucleotides. Substitutions in the DNA nucleotide sequence of a structural gene may result in changes in the amino acid sequence of the polypeptide encoded by the gene, although this is not always the case because of the degeneracy of the genetic code. Consider the DNA triplet TCG (corresponding to AGC in messenger RNA) which codes for the amino acid serine. If the third nucleotide changes to A (U in mRNA), the triplet will still code for serine; but if it changes to C, the resulting codon (TCC in DNA, AGG in mRNA) will instead code for arginine.

In a structural gene, a nucleotide substitution that does not change the amino acid sequence of the encoded polypeptide may have little or no effect on the organism. A nucleotide substitution that changes the amino acid sequence of the corresponding polypeptide will have greater or lesser effect on the organism depending on whether the biological function of the protein is severely affected. Nucleotide substitutions that change a triplet coding for an amino acid into a terminating triplet are likely to have severe effects because polypeptide synthesis will be stopped at that codon. Point mutations that result in the replacement of one amino acid for a different one are called *missense* mutations. When a triplet coding for an amino acid changes to a terminating codon, the change is called a *nonsense* mutation.

Additions or deletions of nucleotide pairs in the DNA sequence of a structural gene result in an altered sequence of amino acids in the encoded polypeptide. From the point of the insertion or deletion, the "reading frame" of the code is shifted, unless the number of additions or deletions equals three or a multiple of three. Additions or deletions of nucleotide pairs in numbers other than three or multiples thereof are called *frameshift* mutations.

Point mutations occur spontaneously, i.e., by naturally occurring causes. The rate of incidence of point mutations may, however, be increased by exposure to high-frequency radiations and by treatment with a variety of chemicals known as mutagens. Chemical mutagens include hydroxylamine, nitrous acid, various alkylating agents such as mustard gas and the epoxides, and certain aromatic compounds known as acridines. The rates of spontaneous mutation vary from organism to organism and from one gene locus to another within the same organism. In viruses, bacteria, and unicellular organisms, recorded mutation rates per gene per cell division range from less than 10^{-9} to 10^{-6}. In higher organisms most mutation rates range from 10^{-6} to 10^{-4} per gene per gamete.

Mutations are rare or ubiquitous events depending on how we choose to look at them. The mutation rates of individual genes are low, but each organism has many genes, and species consist of many individuals. Assume that a typical multicellular organism has 50,000 pairs of genes and that the average rate of mutation per gene is 10^{-5}. On the average, each individual would have $2 \times 50{,}000 \times 10^{-5} =$ one new mutation. Even in a given single gene locus, the incidence of new mutations is high when entire species are considered. The median number of individuals per insect species is estimated to be greater than 10^8. If the average mutation rate per gene per generation is 10^{-5}, more than $2 \times 10^8 \times 10^{-5} = 2{,}000$ new mutations would appear on the average at each locus every generation. Other organisms may consist of fewer individuals per species than insects, but even so a large number of new mutations are likely to arise every generation. The potential of the mutation process to generate new genetic variation is indeed great.

EVOLUTION OF GENOME SIZE

Evolutionary change occurs not only through point mutations but also by changes in the amount and organization of the genetic materials. Mirsky and Ris (1951) were the first to measure the amount of DNA in the cell nucleus of a variety of organisms, including various invertebrate groups as well as fish, amphibians, reptiles, birds, and mammals. At present the DNA content per cell has been measured in more than 1,000 species, including nearly 300 species of fish, more than 100 species of molluscs, and more than 800 plant species. The patterns and adaptive significance of variation in genome size are discussed in Chapter 11.

Evolutionary changes in the amount of DNA may be due to a variety of processes, including polyploidy, polyteny, duplications, and deletions. Polyploidy is a common phenomenon in the evolution of most groups of plants, but it is a rare phenomenon in animals because polyploidy cannot become easily established in species with separate sexes and regular outcrossing. Polyploid species occur among hermaphroditic animals, such as earthworms and planarians, and among animals with parthenogenetic reproduction, including some beetles, moths, sow bugs, shrimps, fish, and salamanders. Polyteny, i.e., multiplication of the number of DNA strands within a chromosome, occurs in certain animal tissues, such as the salivary glands, gut, fat bodies, ovarian nurse cells, and Malpighian tubules of Diptera, but several sources of evidence indicate that polyteny is not a general phenomenon responsible for evolutionary increases of DNA (Rees, 1974; Bachmann *et al.,* 1974).

Deletions and duplications of relatively small DNA segments appear to be the most general processes by which most evolutionary changes in genome size have taken place. Bachmann, Goin, and Goin (1974) observed that when the genome sizes of many teleost, anuran, or placental species are arranged in a frequency diagram, they conform to a logarithmic normal distribution around a single mode. This indicates that evolutionary changes in genome size are numerous and individually small, as would be the case with duplications and deletions, rather than large and discontinuous, as would be expected if they were due to polyploidy and polyteny.

The ancestral form(s) of life of all DNA-containing organisms probably had a short DNA double helix consisting of only one or a few genes. The tens of thousands of different DNA genes found in the human genotype are descendants of that ancestral short segment through multiple duplications and gradual modification of the nucleotide sequences. Duplications of genetic material, followed by divergence of the duplicated segments toward fulfilling different functions, have indeed played a major role in the evolution of life.

Three general classes of DNA duplications may be distinguished for convenience of discussion. First, there are duplications of single structural genes followed by divergent evolution of the duplicated segments toward fulfilling different functions. Second, there are genes which exist in several copies within each genome but which remain essentially identical to each other in DNA sequence and in function. The presence of several copies of a single gene allows the organism to obtain large amounts of the gene product in short time intervals. Third, there are in eukaryotes relatively short sequences of DNA of various lengths that are each repeated many times, some as many as a million times or more per genome, although not all copies may be exactly identical. These highly repetitive sequences of DNA may be involved in the regulation of gene activity; they are the subject of Chapter 12 and will not be further discussed here.

GENE DUPLICATIONS

One example of the first class of gene duplication (i.e., duplications of structural genes followed by gradual divergence toward fulfillment of different but related functions) is provided by the evolution of the globins. The genes coding for myoglobin and the hemoglobins in the vertebrates can be traced to a single gene that became duplicated some 650 million years (MY) ago. Myoglobins as well as hemoglobins are involved in oxygen transport, myoglobins in muscle and hemoglobins in blood. Myoglobin consists of a single polypeptide chain arranged in a complex three-dimensional structure having in the center a heme group—a protoporphyrin ring with an iron atom. Myoglobin has a molecular weight of about 17,000. Most hemoglobin molecules consist of four subunits: two polypeptide chains of one kind and two of another kind, each with a heme group in its center. With the exception of certain Cyclostomata (fish without jaws, such as lampreys and hagfish), all vertebrate hemoglobins are apparently made up that way. Hemoglobin molecules have a molecular weight of about 67,000, approximately 4 times the size of myoglobin. In normal human adults, there are two types of hemoglobins, A and A_2. Hemoglobin A, the most common, consists of two alpha and two beta polypeptide chains (symbolized as $\alpha_2\beta_2$); hemoglobin A_2 consists of two alpha and two delta chains ($\alpha_2\delta_2$). A different hemoglobin is found in human embryos: fetal hemoglobin consisting of two alpha and two gamma chains ($\alpha_2\gamma_2$). The alpha, beta, gamma, and delta polypeptides are each coded by a different gene.

The phylogeny of the hemoglobin genes is shown in Figure 3. (See also Chapter 9.) The gene ancestral to the modern hemoglobin genes became duplicated some 380 MY ago; one gene specialized in coding for alpha-like chains, the other was more similar to present-day beta chains. This duplication made possible the development of the tetramer structure consisting of two copies of each of two polypeptides that is found in the hemoglobins of the higher vertebrates. This structure enhances the possibility of heme-heme interactions that result in more efficient oxygenation and deoxygenation. Another duplication occurred around 150 MY ago, eventually giving rise to the genes coding for the embryonic gamma and the adult beta chains. One more duplication occurred some 35 MY ago in the ancestral lineage of the higher primates—one of the genes coded for the beta chain, the other evolved into the gene coding for the delta chain (Zukerkandl and Pauling, 1965; Zuckerkandl, 1965). The delta chain made possible the appearance of hemoglobin A_2 in adult humans and anthropoid apes. (The beta-delta duplication must have occurred after the evolutionary divergence of the hominoids from the other primates, because the lower primates do not have hemoglobin A_2.)

Other instances of gene duplications are known, for example, the genes coding for trypsin and chymotrypsin well studied in cattle (Neurath *et al.,* 1967) and the genes *PGI–1* and *PGI–2,* which code for the enzyme phosphoglucose isomerase in bony fishes (Avise and Kitto, 1973; for plants see Chapter 8). There is evidence, however, suggesting that most structural genes (i.e., genes translated into proteins) exist in each genome in single copies (Goldberg *et al.,* 1973; Davidson and Britten, 1973). This implies that although gene duplications may not be rare events on the evolutionary time scale, they are not continuously becoming established. In general, enough time passes between duplications of a given gene to allow for the evolutionary divergence of the duplicated genes through point mutations.

The second class of duplicated DNA sequences involves genes repeated from a few to several hundred times in each genome, all copies of a given gene being identical or virtually so. This class of duplicated genes includes such genes as those coding for ribosomal RNA and for transfer RNA, which are transcribed but not translated, and the genes coding for histone proteins, which are transcribed and translated.

Ribosomes are involved in protein synthesis. They become associated with mRNA and mediate the sequential codon-anticodon recognition between mRNA and the tRNAs; the ribosomes make possible the formation of peptide bonds between the amino acids brought in by the tRNAs. There are three kinds of rRNA, which in eukaryotes are designated 5S,

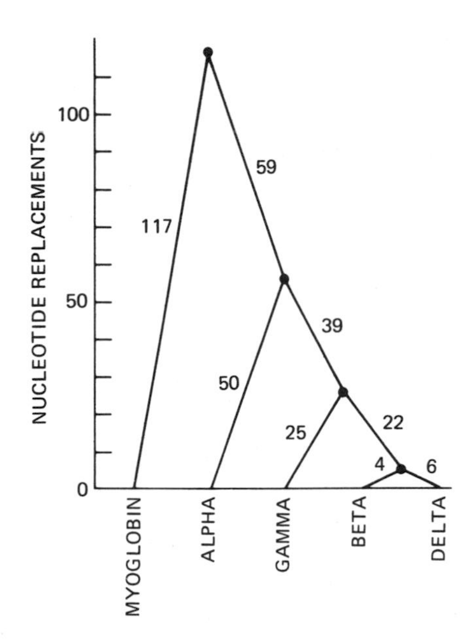

FIGURE 3. Phylogeny of the globin genes. The black dots indicate where ancestral genes were presumably duplicated giving rise to a new gene line. Numbers on segments are the nucleotide replacements required to account for the descent of the genes coding for the five globin molecules from a common ancestral gene. (After Fitch and Margoliash, 1970.)

18S, and 28S, the larger numbers indicating larger RNA molecules. The ribosomes consist of two subunits, one small and one large, made up of the three kinds of rRNA and some 50 kinds of protein. The small ribosomal subunit contains an 18S rRNA molecule; the larger subunit contains a 5S and a 28S rRNA molecule.

In eukaryotes the genes coding for the 18S and 28S rRNA are located in the nucleolar organizer (NO), a chromosomal region lying in the nucleolus. Each gene codes for a single RNA sequence, which after transcription is split into 18S and 28S. The genes coding for the 18S and 28S rRNA are replicated a variable number of times in different organisms. In *Drosophila melanogaster* the NO contains about 130 sequences of this gene (Ritossa and Spiegelman, 1965; Ritossa *et al.,* 1966); in the African toad, *Xenopus laevis,* each NO carries about 400 repeated sequences of this rRNA gene (Brown and Gurdon, 1964). The repeated copies of this rRNA gene are arranged in tandem, although they are separated by short interspersed (spacer) sequences of DNA (Davidson and Britten, 1973). The genes coding for the 5S rRNA are also replicated many times in eukaryotes, although they are not located next to the genes coding for 18S and 28S rRNA. In *D. melanogaster* the NO is in the X and Y chromosomes, while the genes coding for 5S rRNA are in one of the autosomes.

Transfer RNA molecules are involved in protein synthesis as the carriers of specific amino acids to the mRNA-polyribosome complexes, where the anticodon site of a tRNA molecule "recognizes" a corresponding codon in mRNA. The active forms of tRNA are relatively small molecules, with about 70 to 80 nucleotides and a molecular weight of about 30,000; but their precursors have about 40 additional nucleotides which are eventually cleaved off to yield the functional tRNA molecules. There are 30 to 40 different tRNA molecules in the bacterium *Escherichia coli* and about 60 in higher organisms. Because only 20 different amino acids are used in protein synthesis, it follows (as stated earlier) that two or more tRNAs may have affinity for a single amino acid. *Escherichia coli* apparently has only a single copy of each of its tRNA genes, but eukaryotes have multiple copies of each. In yeast there are about 400 tRNA genes per haploid genome, each of about 61 different genes being repeated five to seven times. In *D. melanogaster* each tRNA gene exists in about 13 copies per genome, with a total of somewhat more than 700 genes.

Histones are basic proteins containing a relatively high proportion of the amino acids lysine and arginine; their molecular weights range from 11,000 to 21,000. There are five types of histones, distinguished by the relative amounts of lysine to arginine. Histones and DNA occur in about equal amounts by weight in the chromosomes of eukaryotes. Apparently most of or all the DNA in the nucleus is associated with a layer of histone forming a nucleohistone fiber. Histones may be involved in gene regulation by altering the transcriptional properties of the DNA. The genes coding for the five histone types exist in the form of tandem clusters of multiple copies of a given gene (Kedes and Birnstiel, 1971; Davidson and Britten, 1973).

REARRANGEMENTS OF THE GENOME

Chromosomal inversions and translocations are rearrangements of the genome without addition or deletion of hereditary materials and without changes in the number of chromosomes. Inversions are 180° rotations of chromosome segments. In inversion heterozygotes, synapsis of the homologous chromosomes requires the formation of a loop involving the inverted segments. Inversions can be recognized by such loops in microscope preparations. The presence of inversions can also be detected because they effectively suppress recombination in heterozygotes. Chromosomal inversions are known to be common evolutionary phenomena in organisms, such as *Drosophila* flies and *Chironomus* midges, that provide favorable materials for their study.

Reciprocal translocations involve the interchange of blocks of genes between nonhomologous chromosomes. In a translocation heterozygote, pairing during meiosis results in a cross-shaped configuration by which the translocation heterozygotes can be recognized. Translocation heterozygotes are usually partially sterile because many of their gametes have some chromosome parts duplicated and others missing and thus cannot result in normal progeny. Translocation heterozygotes are very rare in animals, but they have been found in natural populations of plants.

Chromosomal fusions and fissions do not change the amount of hereditary material, but they do alter the number of chromosomes. Fusions and fissions are common evolutionary phenomena. The haploid chromosome numbers in most animals lie between six and 20, but the range extends from one in the nematode *Parascaris equorum* var. *univalens* to about 220 in the butterfly *Lysandra atlantica.* In plants the gametic number of chromosomes may be higher than 600, as in the fern *Ophioglossum petiolatum,* which is almost certainly a polyploid.

Recent cytogenetic techniques, such as quinacrine fluorescent staining, have made it possible to make detailed comparisons of the chromosome structure of related species. Man has 46 chromosomes; its closest animal relative, the chimpanzee (*Pan troglodytes*), has 48 chromosomes. Thus, at least one chromosomal fusion or one fission has occurred in the evolution of man and chimpanzee from their most recent common ancestor. Other chromosomal rearrangements have also been detected, including some translocations (e.g., the long arm of chromosome 9 in chimpanzee is homologous to the long arm of chromosome 5 in man) and inversions, although overall the chromosomes of man and chimpanzee remain fairly similar (Lin *et al.,* 1973). Chromosomal rearrangements change the linkage relationships between genes and may also affect gene regulation. (See Chapter 13.)

GENETIC VARIATION AND EVOLUTION

Evolutionary change occurs in populations, not in individuals. The individual is born, grows, and eventually dies. From the evolutionary point of view the individual is ephemeral; only the population is continuous; the continuity derives from the mechanism of biological heredity. Individual organisms may change throughout their lifetimes, but their genetic constitution remains constant. On the contrary, the genetic constitution of a population may change from one generation to another, and it usually does. Evolution consists of changes in the genetic constitution of populations.

Genetic change in populations occurs through the processes of mutation, migration, random drift, and natural selection. Mutation is the ultimate source of all genetic variation. Migration or gene flow from one population to another of the same species may increase genetic variation in local populations. Random genetic drift is due to sampling errors from one generation to another. Because all populations are finite, the genetic frequencies change from generation to generation because of accidents of sampling; the smaller the population, the greater the effect of random drift on genetic frequencies. Natural selection may be simply defined as differential reproduction of alternative genetic variants. Organisms having genetic variations useful to their carriers as adaptations to the environment are likely to leave more progeny than organisms with alternative, less adaptive genetic variations. The more adaptive variations will increase in frequency through the generations, while the less adaptive variations will decrease. Mutation, migration, and drift are random processes with respect to adaptation; genetic variants increase or decrease in frequency by these processes independently of whether they are useful to their carriers as adaptations. Natural selection is the only evolutionary process that is directional with respect to adaptation. The adaptive nature of organisms and of their structures, physiology, and behavior is due to natural selection.

The evolutionary potential of a population is determined by the degree of genetic variation in the population. Natural selection, as well as random drift, can only take place in populations that possess genetic variability. The more genetic variation there is in a population, the greater the opportunity for the operation of natural selection. The correlation between genetic variation and rate of evolutionary change by natural selection is expressed in the Fundamental Theorem of Natural Selection (Fisher, 1930): "The rate of increase in fitness of a population at any time is equal to its genetic variance in fitness at that time."

Ayala (1965, 1968) has provided experimental evidence of the positive correlation between amount of genetic variation and rate of evolutionary change. Laboratory populations were set up with *Drosophila serrata* flies collected in various localities. Two types of experimental populations were established: "Single-strain" populations had founders descended from flies collected in a single locality; "mixed" populations were started by crossing flies collected in two different localities. The adaptation of a population to the experimental environment was measured by the number of individuals in a population. All populations gradually increased their adaptation to the experimental environment. However, "mixed" populations increased in numbers about twice as fast as the "single-strain" populations (Figure 4). The "mixed" population had from the start about as much genetic variation as the two "single-strain" parental populations together, because it was started by mixing their gene pools. "Mixed" populations evolved at a much faster rate than either one of the parental populations.

Other experiments have shown that under stringent conditions of natural selection populations in which the genetic variation was increased by ionizing radiations may evolve at a faster rate than nonirradiated, genetically less variable populations (Ayala, 1966, 1969).

The amount of genetic variation in a population is a fundamental

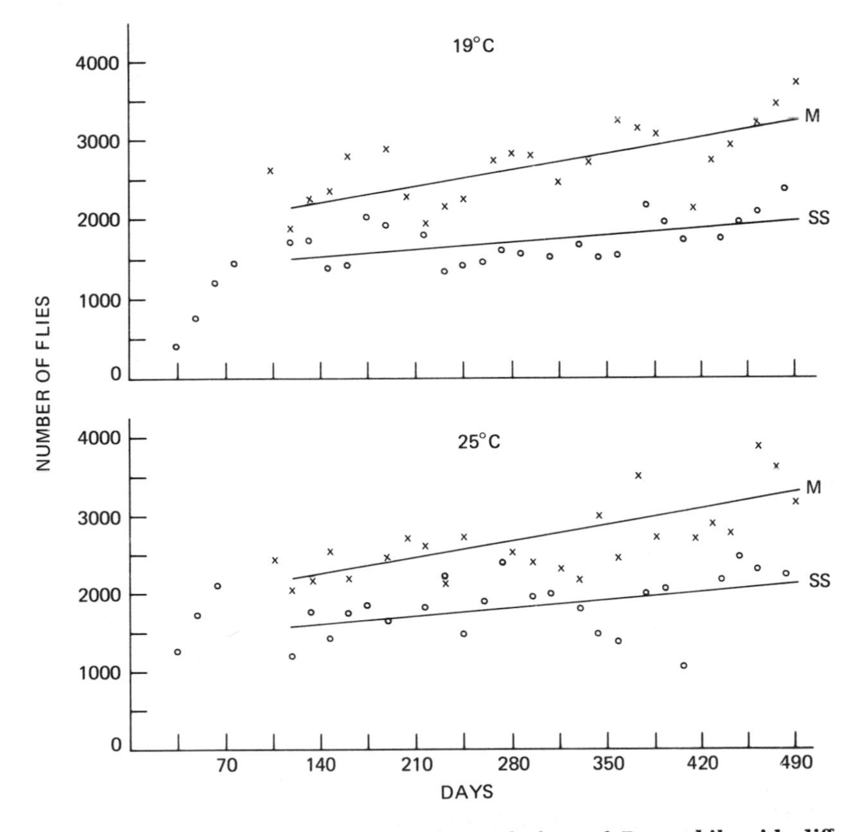

FIGURE 4. Evolution of experimental populations of *Drosophila* with different levels of genetic variation. (After Ayala, 1965.) The average number of flies increases with time in all populations as a consequence of the gradual adaptation of the populations to the experimental environment. The rate of evolution is, however, greater in populations having more genetic variation ("mixed" populations, M) than in those with lesser genetic variation ("single-strain" populations, SS). The experiments were performed at two temperatures, 19 and 25°C, with similar results. The populations were studied for 500 days, or about 25 generations.

parameter in evolutionary studies, because genetic variation determines the evolutionary potential of populations. Genetic polymorphisms are pervasive in natural populations, as has been shown by studies of morphological variation, artificial selection, fitness modifications, and other methods (review in Lewontin, 1974). Recently, the techniques of gel electrophoresis and selective assay of enzymes have made it possible to estimate, at least to a first approximation, the proportion of all genes that are polymorphic in a population and how polymorphic they are. A great deal of information has accumulated in the last decade showing that at least 30 to 80 percent of all structural gene loci are polymorphic in natural populations of most organisms and that an average individual is heterozygous at least in 5 to 20 percent of its structural genes. This information is reviewed from various points of view in Chapters 2 to 5. These chapters also discuss the processes maintaining genetic variation.

ANAGENESIS AND CLADOGENESIS

The process of evolution may be seen in two ways, which have been called *anagenesis* and *cladogenesis* (Rensch, 1960). Anagenesis, or phyletic evolution, consists of changes occurring within a given phylogenetic lineage as time proceeds. The stupendous changes from a primitive form of life some 3 billion years ago to man, or to some other modern form of life, are anagenetic evolution. Cladogenesis occurs when a phylogenetic lineage splits into two or more indepedently evolving lineages. The great diversity of the living world is the result of cladogenetic evolution. The evolutionary geneticist is interested in the amount of genetic differentiation that takes place during anagenetic and cladogenetic events.

Among cladogenetic processes, the most decisive one is speciation—the process by which one species splits into two or more species. Species are generally defined as arrays of populations that are reproductively isolated from any other such arrays. Species are, therefore, independent evolutionary units. Adaptive changes occurring in an individual or population may be extended to all members of the species by natural selection; they cannot, however, be passed on to different species. A basic question with respect to cladogenesis is how much genetic differentiation accompanies the formation of new species. Quantitative answers to this question have become possible only in recent years through the application of molecular techniques such as gel electrophoresis. The speciation process and the degree of genetic change concomitant with speciation are discussed in Chapters 6 to 8.

Anagenetic changes in morphology may be ascertained whenever a fossil record exists for a given lineage. It might seem that there is no way to determine anagenetic change at the level of the DNA, proteins, and other molecules. These molecules are not as a rule preserved in ancestral organisms in such a way that their composition and structure can be compared with the molecules of living organisms. There is, however, an indirect way to estimate the amount of anagenetic change at the molecular level. Assume that A is the most recent common ancestral species of two closely related living species, B and C. Assume now that we compare the primary structure of a given protein (or the DNA sequence of the gene coding for it) in B and C, and that the number of differences between them is x. If we further assume that no convergent or parallel evolutionary changes occur at the molecular level, we may conclude to a first approximation that the amount of anagenetic change between *A* and *B*, or between *A* and *C*, is simply $x/2$. This simple rationale is used in Chapters 9 and 10 to estimate molecular anagenetic evolution. In fact, when several living species are studied, the assumption that the amount of anagenetic evolution is the same in the two legs of a phylogenetic bifurcation can be removed. Moreover, the possibilities of convergent and parallel evolution can be corrected for.

Evolution is, on the whole, a gradual process of change. The relative degree of similarity among living species has been traditionally used as a measure of recency of common ancestry in studies of comparative anatomy, embryology, ethology, and other branches of biology. Relative degrees of similarity in molecular composition may also be used to infer phylogenetic relationships. Gel electrophoresis, immunology, amino acid sequencing of proteins, DNA hybridization, and other molecular techniques have in recent years greatly contributed to the reconstruction of phylogenetic history. The primary structure of proteins, in particular, contains a great deal of phylogenetic information. (See Chapters 9 and 10.) Protein sequencing and other molecular methods may, in fact, become in the near future the most powerful tools for the study of phylogeny.

The last three chapters of this book are primarily concerned with evolution at the level of the DNA. Chapter 11 deals with the evolution of genome size—overall changes in the amount of DNA per cell. Chapter 12 considers the evolution of DNA sequences that are found in multiple copies in the genome of a given organism. Chapter 13 contrasts the evolution of structural genes with the evolution of gene regulation, and their relative roles in the history of life.

SUGGESTED READINGS

The basic concepts of molecular genetics introduced in Chapter 1 are discussed in any modern textbook of genetics. An excellent one is U. Goodenough and R. P. Levine, *Genetics,* Holt, Rinehart and Winston, New York, 1974.

Readers willing to study molecular genetics in depth may use as guides G. S. Stent, *Molecular Genetics,* W. H. Freeman, San Francisco, 1971; and J. D. Watson, *Molecular Biology of the Gene,* third ed., W. A. Benjamin, Menlo Park, California, 1975.

The best single-volume treatment of population and evolutionary genetics is T. Dobzhansky, *Genetics of the Evolutionary Process,* Columbia University Press, New York, 1970. Another excellent book is E. Mayr,

Population, Species, and Evolution, Belknap Press, Cambridge, Massachusetts, 1970. A brief introduction to population biology is E. O. Wilson and W. H. Bossert, *A Primer of Population Biology,* Sinauer Associates, Sunderland, Massachusetts, 1971.

The current literature in ribosomal RNA and histone genes is briefly reviewed in M. L. Pardue, Repeated DNA sequences in the chromosomes of higher organisms, *Genetics,* **79** (suppl.), 159–170, 1975.

A readable account of the evolution of the globins is E. Zuckerkandl, The evolution of hemoglobin, *Sci. Amer.,* **212**(5), 110–118, 1965; for a more complete review see V. M. Ingram, *The Hemoglobins in Genetics and Evolution,* Columbia University Press, New York, 1963. A general but uneven review of gene duplications can be found in S. Ohno, *Evolution by Gene Duplication,* Springer-Verlag, New York, 1970.

CHAPTER 2

GENIC VARIATION IN NATURAL POPULATIONS

ROBERT K. SELANDER

By combining electrophoresis with histochemical staining methods, Hunter and Markert (1957) developed the zymogram technique for identifying enzymes in tissue extracts. Its application soon led to the important discovery of isozymes, or multiple molecular forms of enzymes (Markert and Møller, 1959). And 10 years ago, the technique was introduced to population biology when Lewontin and Hubby (1966) and Harris (1966) systematically attempted to measure genic variation in natural populations by screening randomly selected samples of proteins for allozymes—genetically controlled variants of isozymes. Since then, increasing numbers of population geneticists and evolutionists have become preoccupied with the problem of the biological meaning of polymorphic variation in the primary structure of proteins. (See Le Cam *et al.,* 1972; Lewontin, 1974; Markert, 1975; and Nei, 1975, for reviews.) As evidence of the nearly universal occurrence of extensive protein polymorphism in natural populations accumulated, views regarding its evolutionary significance soon became polarized. Some workers believed that most molecular variation would prove to be physiologically meaningful and hence, under selective control and important in adaptation; but others regarded it as evolutionary "noise," without phenotypic effect and thus selectively neutral. The various chapters in this book will demonstrate that the selectionist versus neutralist controversy continues to dominate the field of molecular evolution and that we are still far from our goal of answering this fundamental question.

The objectives of this chapter are to summarize information on the extent and major dimensions of polymorphic variation in proteins in natural populations of organisms and to outline some major aspects of current theory relating to the evolutionary significance of this variation, particularly modifications of the neutral theory that have developed since Lewontin's (1974) comprehensive review of the field. Some aspects of molecular variation, including possible adaptive patterns and relationships to enzyme function, are treated at greater length in Chapters 3 to 5.

EXTENT AND DIMENSIONS OF PROTEIN POLYMORPHISM

According to recent estimates, less than 10 percent of the DNA in eukaryotic genomes involves genes that are transcribed and translated to produce functional enzymes and other proteins, the number of which is on the order of 40,000. (See Chapter 12.) Because we cannot directly measure individual gene variation in the DNA, estimates of genic variation in populations are based on assessments of heterogeneity in the structural gene products, that is, enzymes and other proteins. Moreover, these analyses utilize indirect measures of structural heterogeneity, because techniques of amino acid sequencing are too time-consuming and expensive to be applied on the large scale required for population genetics.

When we consider that approximately 30 percent of DNA base changes cause no modification in the amino acid sequence of proteins because of the redundancy of the genetic code and that most amino acid substitutions do not alter the electrostatic charge of proteins and are therefore undetectable by electrophoresis, we can appreciate the severe limitations of available techniques for studying molecular variation in populations.

ELECTROPHORETIC TECHNIQUES

When proteins in blood or extracts of other tissues or of whole organisms are allowed to migrate in a supporting medium (for example, a slab of starch or acrylamide gel) under the influence of an electric field, differences in net electrostatic charge are reflected in different mobilities. Following electrophoresis, the positions of specific enzymes, such as glucose-6-phosphate dehydrogenase, or of nonenzymatic proteins (e.g., albumin) can be demonstrated by specific histochemical stains applied directly to the electrophoretic medium (Figure 1). The resulting phenotypes are bands of stain indicating regions of enzyme activity or protein concentration for which the term *electromorph* recently has been proposed (King and Ohta, 1975).

Although a mobility difference between polypeptides generally can be taken as evidence of at least one amino acid difference, identity of mobility does not indicate identical amino acid sequence. This is because

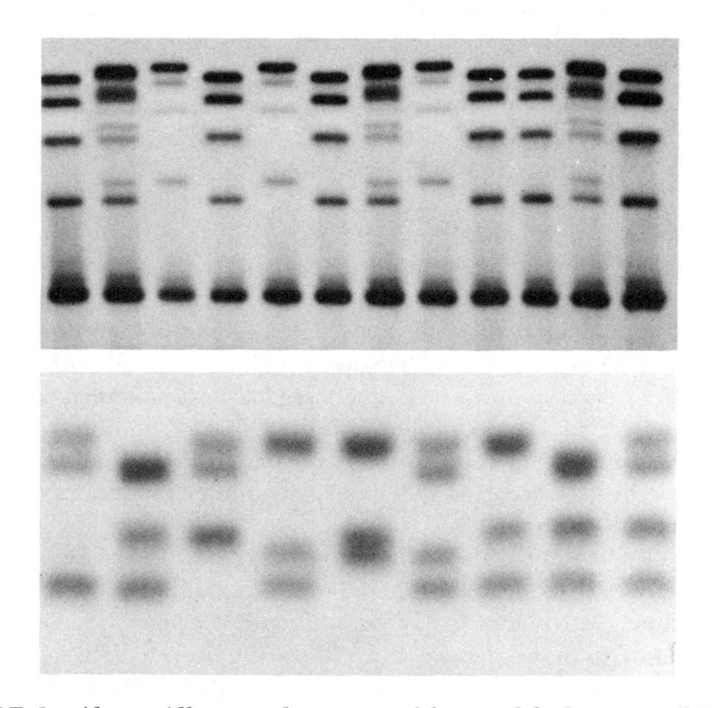

FIGURE 1. Above: Allozyme phenotypes of lactate dehydrogenase (LDH) in kidney extracts of the old-field mouse (*Peromyscus polionotus*). LDH is a tetramer of polypeptides encoded by two loci (*Ldh-1* and *Ldh-2*). In individuals homozygous at both loci, the polypeptides combine to produce a five-banded pattern; in heterozygotes a pattern of 15 bands is produced, not all of which are resolved. In this sample of 12 individuals, the *Ldh-1* (upper system) is polymorphic for fast- and slow-migrating allozymes (F and S, respectively); phenotypes (and presumed genotypes) of individuals from left to right are SS, FS, FF, SS, FF, SS, FS, FF, SS, SS, FS, and SS. *Ldh-2* (lower system) is monomorphic. Below: Variation in two isozymes of leucine amino-peptidase (LAP) in the brown snail (*Helix aspersa*). LAP is a monomer; homozygotes are single-banded, and heterozygotes have two bands of equal intensity. The upper system (*Lap-1*) is polymorphic for two allozymes, and the lower system is polymorphic for three allozymes. Phenotypes of the first five individuals (from left to right) are as follows: *Lap-1,* FS, SS, FS, FF, and FF; *Lap-2,* SS, FS, FF, MS, and FM. M symbolizes an allozyme with intermediate migration.)

16 of the 20 common amino acids have nonionizable side chains and are electrophoretically neutral in the pH range of buffers generally employed in electrophoretic studies. Only four amino acids are charged: arginine and lysine are basic (negative), and glutamic acid and aspartic acid are acidic (positive). The mobility of a protein depends primarily

on the net charge of the amino acids on the surface; in globular proteins the basic and acidic ones tend to occur on the surface, while the uncharged, hydrophobic ones are in interior positions (Dickerson and Geis, 1969). Mobility is also determined to some degree by the conformation (folding pattern) of the molecule.

At the pH values normally used in electrophoretic studies, the proportion of accepted point mutations that would be detectable electrophoretically is estimated at 0.27 (King and Wilson, 1975). However, this value is not expected to be constant over all molecules, and there is evidence that it is not (Johnson, 1974*a;* King, 1973). Still, it is noteworthy that by using 0.27 as the proportion of electrophoretically detectable substitutions King and Wilson (1975) estimated for a comparison of chimpanzee and man that the expected number of amino acid differences per protein is 8.2 per 1,000 sites, which agrees well with a value of 7.2 obtained from amino acid sequencing or microcomplement fixation analyses of 12 of the 40 proteins studied. An assumption underlying this calculation is that substitutions at a particular amino acid site of a given protein have occurred independently and at random in the human and chimpanzee lineages.

Whatever the mean and variance of the proportion of detectable substitutions over diverse proteins may be, we know that proteins identified as the same electromorph may be sequentially heterogeneous because much if not most of the sequence variation is not detectable by the zymogram method. Although we translate electromorph frequencies into allele frequencies and speak of genotypes corresponding to electromorph patterns, it is with the understanding that the "alleles" we designate may actually be groups of isoalleles.

The discriminative power of electrophoretic methods varies greatly for different proteins for several reasons, including the tendency of some enzymes to band more tightly than others (Harris *et al.,* 1974). Variation among researchers in the quality of application of the zymogram technique and in standards of interpretation of electromorph patterns also must be considered in evaluating reports on genic variation in populations.

In preparing this chapter I assembled and analyzed estimates of heterozygosity in approximately 140 species of animals and plants. The data are derived from electrophoretic surveys in which some attempt was made to sample structural gene loci without prior knowledge of whether they were monomorphic or polymorphic in the organisms studied.

ESTIMATING GENIC VARIATION

The results of surveys attempting to measure variation at structural gene loci are discussed in detail beyond. Here it will suffice to note that in many organisms populations are polymorphic (for electromorphs) at 25 to 50 percent of their loci (P = 0.25 to 0.50) and individuals are on the average heterozygous at 5 to 15 percent of their loci. (Except as noted, this chapter deals with estimates of variation in continental species having large ranges and, presumably, large effective population sizes.) A locus is here defined as polymorphic in a population if the frequency of the commonest electromorph (allele) is equal to or less than 0.99. Heterozygosity for a locus (h) is defined as $1 - \Sigma X_i^2$, where X_i is the frequency of the ith allele, and average heterozygosity (H) is the mean of h over all loci examined. For organisms having nonrandom mating populations, such as self-fertilizing plants and snails, H is a useful measure of genic variability, or gene diversity (Nei, 1975), although it is not related to the frequency of heterozygotes in populations. Standard errors of H are very large because the variance of h includes a considerable interlocus component. Several measures of genic variation other than P and H currently in use, including the effective number of alleles per locus (the reciprocal of homozygosity), are less satisfactory for most purposes (Nei, 1975).

To illustrate the population geneticist's approach in estimating genic heterozygosity from molecular data, we will consider the results of research on human populations (Table I). In a survey of electrophoretically detectable variation at 71 enzyme loci in European populations, Harris and Hopkinson (1972) found a polymorphic proportion of 0.28 and an average heterozygosity of 0.067. Recently, Nei and Roychoud-

TABLE I. Genic variation in human populations*

Race and number of loci	Proportion of loci: Polymorphic	Proportion of loci: Heterozygous (± S.E.)	Codon difference: D_z	Codon difference: D_z'
Caucasoid				
A. 74	0.31	0.099 ± 0.021	0.104	0.130
B. 62	0.32	0.104 ± 0.023	0.110	0.137
C. 35	0.40	0.142 ± 0.034	0.153	0.187
D. 71	0.28	0.067		
Negroid				
B. 62	0.40	0.092 ± 0.019	0.097	0.115
C. 35	0.51	0.122 ± 0.028	0.131	0.151
Mongoloid				
C. 35	0.40	0.098 ± 0.027	0.103	0.122

A: All loci for Caucasoids; B: common loci for Caucasoids and Negroids; C: common loci for all three races; and D: estimates by Harris and Hopkinson (1972), based on 71 enzyme loci.

*After Nei and Roychoudhury (1974).

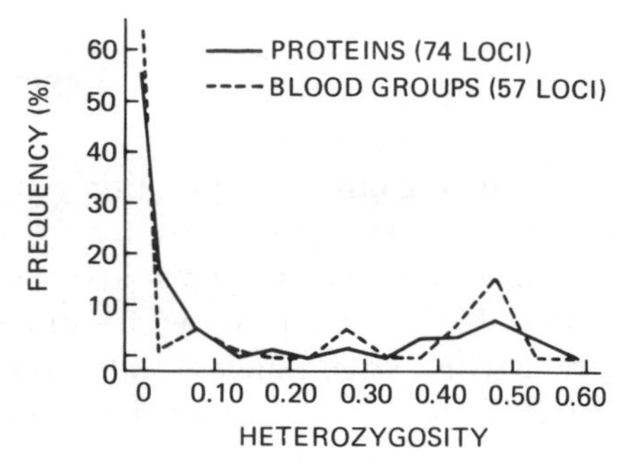

FIGURE 2. Frequency distributions of heterozygosity (*h*) for protein and blood group loci in man (Caucasoids). (From Nei and Roychoudhury, 1974.)

hury (1974) have analyzed variation in the three major races of man, Caucasoids, Negroids, and Mongoloids, using allele frequency data pertaining to 74, 62, and 35 structural gene loci, respectively. Their values of P and H are larger than those reported by Harris and Hopkinson (1972) largely because of the inclusion of 12 nonenzymatic proteins, nine of which are highly polymorphic (for example, α-acid glycoprotein, with $h = 0.394$ to 0.484 in the races) and have been intensively studied by human geneticists for just that reason. With nonenzymatic proteins omitted from the samples, average heterozygosity reduces to about 0.084, which is not significantly different from the value obtained by Harris and Hopkinson (1972).

There are no significant racial differences in average heterozygosity or proportion of loci polymorphic. Frequency distributions of h are similar in all three races, with about 70 to 80 percent of loci having a heterozygosity lower than 0.10 (Figure 2). Correlations of heterozygosity between pairs of races are high, ranging from $r = 0.72$ for the Negroid-Mongoloid to 0.86 for the Caucasoid-Mongoloid comparisons.

Table I includes standard (D_x and maximum $D_{x'}$) estimates of codon differences per locus between two randomly chosen genomes. Because these estimates are only slightly larger than average heterozygosity (a minimum estimate of codon differences), Nei and Roychoudhury (1974) concluded that differences between alleles are in most cases caused by single codon changes.

By screening a quarter of a million Europeans for rare alleles at 43 enzyme loci, Harris *et al.* (1974) determined that, on average for any locus, between one and two individuals per 1,000 are heterozygous for a rare allele (frequencies less than 0.005) determining an electrophoretic variant. The proportion of these mutations arising in the germ line of one of the parents of variant individuals was estimated at less than 2 percent. If the unusually variable enzyme placental alkaline phosphatase is excluded from consideration, polymorphic and monomorphic loci do not differ in the incidence of rare alleles.

Analyses of frequencies of red blood cell antigens in man have yielded estimates of average heterozygosity similar to those for protein loci (Lewontin, 1967; Nei and Roychoudhury, 1974). However, Nei (1975) notes that gene diversity at the codon level may not be the same for the two kinds of loci because the relationship between the immunological reaction and the gene is still not well understood. Data for white cell antigens (histocompatibility loci) are similarly difficult to interpret because of questions regarding allelism.

PROTEIN VARIATION IN RELATION TO METABOLIC FUNCTION

The earliest surveys indicated that some proteins are more likely to be polymorphic than others. The extent of this heterogeneity is enormous, as shown for *Drosophila* and rodents in Figure 3. Some proteins are almost universally variable in populations, while others rarely if ever are polymorphic. To account for this dimension of variation, Gillespie and Kojima (1968) on the basis of work with *Drosophila* suggested that the degree of polymorphism in enzymes varies according to function, being generally lower in glucose-metabolizing enzymes (group I) than in nonspecific enzymes (group II). Subsequently, these workers (Kojima *et al.,* 1970) proposed that the greater average variability of group II enzymes can be attributed to a greater diversity of substrates, many of which originate in the external environment. Johnson (1973) also analyzed enzyme variation in *Drosophila* in terms of the utilization of internal versus external substrates and found a larger number of alleles at loci encoding group II enzymes.

Gillespie and Langley (1974) recently have redefined group I enzymes as those "characterized by a singular physiological substrate which is generally generated and utilized intracellularly," and group II enzymes as those "with multiple physiological substrates which reflect environmental diversity." Although their analysis was limited to enzymes, they would classify in group I structural, ribosomal, and regulatory proteins, together with most enzymes involved in biosynthesis, intermediary metabolism, glycolysis, and the citric acid cycle. Gillespie and Langley (1974) believe that the actual proportion of genes in the genome coding for group II enzymes is small and disproportionately represented in surveys of protein variation.

The distinction between group I and group II enzymes may prove to be useful, but present classifications of enzymes according to functional type are unsatisfactory. For example, xanthine dehydrogenase of *Drosophila* is assigned to group I but might justifiably be placed in group II because it has multiple substrates (Glassman, 1965). More importantly, many types of enzymes assigned to either group are heterogeneous with

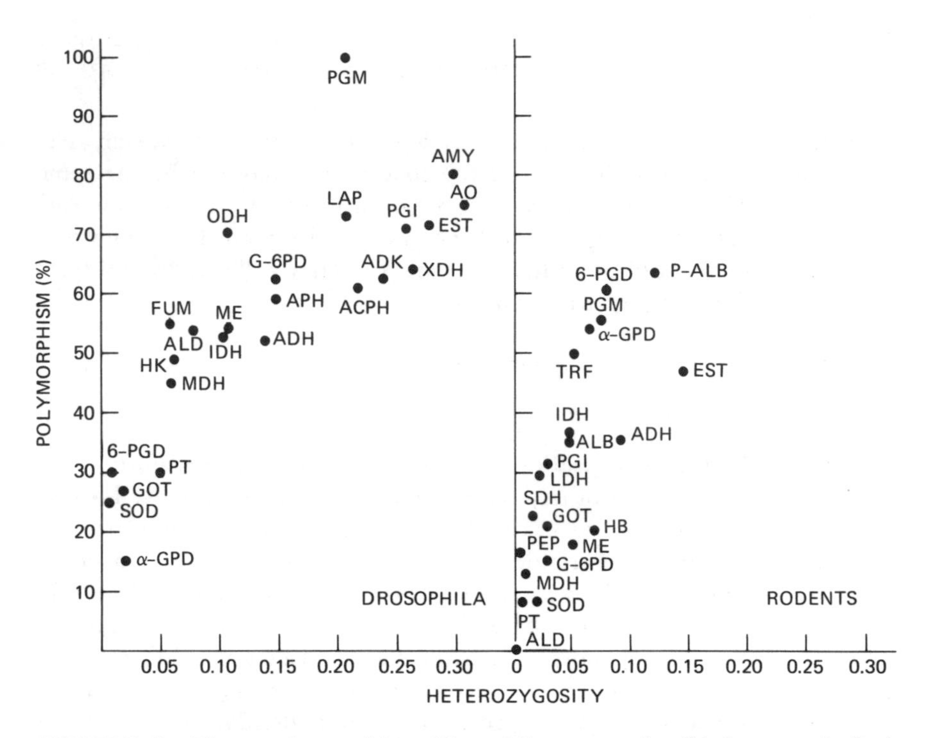

FIGURE 3. Mean polymorphism (*P*) and heterozygosity (*h*) for protein loci in *Drosophila* and rodents. Acid phosphatase (ACPH), alcohol dehydrogenase (ADH), adenylate kinase (ADK), albumin (ALB), aldolase (ALD), amylase (AMY), aldehyde oxidase (AO), alkaline phosphatase (APH), esterase (EST), fumarase (FUM), α-glycerophosphate dehydrogenase (α-GPD), glucose-6-phosphate dehydrogenase (G-6PD), glutamic oxaloacetic transaminase (GOT), hemoglobin (HB), hexokinase (HK), isocitrate dehydrogenase (IDH), leucine aminopeptidase (LAP), lactate dehydrogenase (LDH), malate dehydrogenase (MDH), malic enzyme (ME), octanol dehydrogenase (ODH), pre-albumin (P-ALB), peptidase (PEP), 6-phosphogluconate dehydrogenase (6-PGD), phosphoglucose isomerase (PGI), phosphoglucomutase (PGM), nonenzymatic protein (PT), sorbitol dehydrogenase (SDH), superoxide dismutase (SOD), transferrin (TRF), and xanthine dehydrogenase (XDH).

respect to variability (Zouros, 1975). Hence, the assignment of enzymes to group I or group II does not take into account the distribution of variation within the enzyme classes.

The proportions of loci in 28 *Drosophila* species having various degrees of heterozygosity are shown in Table II. There is in many cases a strongly bimodal distribution. In *Drosophila,* Zouros (1975) found that 23 percent of hydrolases (esterases, phosphatases, peptidases, and amylases) are monomorphic, being among the more conservative loci studied. They are placed in group II (supposedly having multiple or variable substrates), although their *in vivo* functions are unknown and

TABLE II. Proportions of loci having indicated level of heterozygosity in 28 species of *Drosophila*

		Heterozygosity (h) (%)						
Protein	**Number of loci**	**0–0.10***	**0.11–0.20**	**0.21–0.30**	**0.31–0.40**	**0.41–0.50**	**0.51–0.60**	**> 0.60**
α-GPD	26	(46) 96		4				
PGM	16	(0) 19	50	6	13	6	6	
XDH	25	(32) 44		12	4	16	20	4
LAP	30	(27) 34	23	10	17	10	3	3
ODH	27	(26) 56	30	11				4
AO	32	(16) 32	9	3	6	25	16	10
EST	102	(27) 42	6	6	10	9	12	16
APH	27	(33) 59	7	7	11	11	4	
Hydrolases†	187	(29) 44	9	7	11	9	8	12
Nonenzymatic	76	(70) 84	4	8		3	1	

*Percentage of monomorphic loci ($h = 0$) in parentheses.
†Esterases, phosphatases, amylase, and leucine aminopeptidase.

may be quite specific. Conversely, esterase A of the olive fly *Dacus oleae,* which is associated with the nervous system and may have a specific, internal substrate, is highly polymorphic ($h = 0.72$, with more than 15 alleles) (Zouros and Krimbas, 1969).

Variation in the proteins of *Drosophila* can be summarized as follows: Among the group I enzymes, α-glycerophosphate dehydrogenase, 6-phosphogluconate dehydrogenase, and glutamic oxaloacetic transaminase are conservative; malic enzyme and five others are moderately variable; and xanthine dehydrogenase, adenylate kinase, phosphoglucose isomerase, and phosphoglucomutase tend to be strongly polymorphic. Phosphoglucomutase is unusual in being almost universally polymorphic in populations but having only a moderate level of heterozygosity. Esterases, aldehyde oxidase, and some other group II enzymes tend to be highly polymorphic, but alkaline phosphatase and alcohol dehydrogenase are only moderately variable. Nonenzymatic proteins are conservative, as is the enzyme superoxide dismutase (indophenoloxidase), the physiological function of which is uncertain (Beckman and Beckman, 1975). Ribosomal proteins apparently are remarkably conservative

TABLE III. Mean heterozygosity of loci for 28 species of *Drosophila* and 26 species of rodents

	Drosophila		**Rodents**	
Protein	**Number of loci**	**Mean heterozygosity**	**Number of loci**	**Mean heterozygosity**
		Group I		
G-6PD	16	0.152	13	0.030
α-GPD	26	0.019	22	0.067
6-PGD	10	0.013	23	0.081
IDH*	19	0.105	23	0.078
MDH	29	0.060	45	0.012
ME	26	0.111	11	0.052
PGI	7	0.262	19	0.032
PGM	16	0.212	34	0.076
ALD	26	0.079	9	0.000
GOT*	11	0.022	19	0.058
Mean		0.1035		0.0486
		Group II		
EST	102	0.284	72	0.146
ADH	27	0.142	14	0.093
		Group III		
Nonenzymatic	76	0.053	151	0.053
		Unknown Function		
SOD	20	0.017	24	0.020

*Supernatant (cytosol) form.

both within and between species of *Drosophila* (Berger and Weber, 1974; Vaslet and Berger, 1975).

Mean individual heterozygosity (H) is in all groups of organisms roughly twice as large for group II enzymes as for those in group I. As shown in Table III for *Drosophila* and rodents, Gillespie and Langley's (1974) suggestion that levels of variability in mammals and *Drosophila* are similar for group I enzymes but different for those in group II (*Drosophila* being much more heterozygous) is not supported. Enzyme for enzyme, there is no difference between *Drosophila* and rodents in degree of difference in heterozygosity levels for group I and group II. (The number of group II enzymes for which adequate data are available is, unfortunately, only two, esterases and alcohol dehydrogenases.) Nonenzymatic proteins (group III) and superoxide dismutase are similarly low in variability in both groups of organisms. Thus, whatever factors influence variation in degree of polymorphism in *Drosophila* and rodents similarly affect both group I and group II enzymes.

Rodents, like other tetrapods, are about half as polymorphic as *Drosophila* over the common enzymes assayed (Table III). Mean heterozygosity values for group I enzymes in *Drosophila* and rodents are uncorrelated ($r = -0.136$). (Over all proteins compared, $r = 0.399$, also nonsignificant.) Glucose-6-phosphate dehydrogenase and phosphoglucose isomerase are highly variable in *Drosophila* but not in rodents, and 6-phosphogluconate dehydrogenase is more variable in rodents than in *Drosophila*. However, the group II enzymes compared are highly variable in both organisms.

When we compare *Drosophila* and other insects, the correlation of heterozygosities is higher; over 19 proteins (including those in Table IV), $r = 0.596$ ($p < 0.01$), and the means are not different. (As noted beyond, there also is a strong correlation for *Drosophila* and haplodiploid insects.) Apparently the distribution of variability is fairly consistent within large taxonomic groups, such as the insects or the mammals, and quite different between them.

In exploring various aspects of the relationship between polymorphism and metabolic function, Johnson (1974*b*) has maintained that regulatory enzymes should for adaptive reasons be more variable than other enzymes. Earlier (1971) he argued that degree of polymorphism in *Drosophila* is related to the equilibrium constant (K) of the reaction catalyzed by an enzyme, but this interpretation was not supported by Ayala and Powell (1972), who considered the hypothesis ill-conceived. More recently, Johnson (1974*b*) has attempted to use other criteria for distinguishing regulatory and nonregulatory classes of enzymes, but his analysis is unconvincing. Reasons for considering xanthine dehydro-

TABLE IV. Mean heterozygosity of loci in insects

Protein	*Drosophila**	Other insects†	Haplodiploid wasps‡
G-6PD	0.152	0.492	0.000
α-GPD	0.019	0.069	0.002
6-PGD	0.013	0.382	0.030
IDH	0.105	0.034	0.020
MDH	0.060	0.012	0.001
HK	0.065	....	0.000
PGI	0.262	0.283	0.103
PGM	0.212	0.279	0.072
GOT	0.022	0.016	....
XDH	0.265	0.249	....
LAP, PEP	0.191	0.171	0.101
EST	0.284	0.284	0.227
SOD	0.017	0.036	0.000
Nonenzymatic	0.053	0.029	0.000
Mean	0.1229	0.1797	0.0463

*Twenty-eight species.
†Four species of three genera *(Philaenus, Gryllus,* and *Magicicada).*
‡Six species of four genera *(Megachile, Nomia, Opius,* and *Polistes).*

genase regulatory are not apparent, and alcohol dehydrogenase and glutamate-pyruvate amino transferase are so classified only because they "occupy metabolic positions of potential regulatory importance." Additionally, Johnson (1974*b*) classifies alcohol dehydrogenase in *Drosophila* as regulatory, although he also maintains that its physiological role is not clear. However, there is in fact considerable evidence that alcohol dehydrogenase functions physiologically in *Drosophila* to break down alcohols in the environment, including ethanol (Briscoe *et al.,* 1975; Vigue and Johnson, 1973; Clarke, 1975), and this clearly is not a regulatory role. Even if accepted as proposed, the regulatory-nonregulatory hypothesis is not convincing when applied to data. For example, the regulatory enzymes hexokinase and malic enzyme are not strongly polymorphic in *Drosophila.*

Zouros (1975) has suggested that the primary structure, rather than the function, of the protein molecule is an important factor determining the extent of polymorphism. Thus, structurally related enzymes will show similar amounts of variation, and enzymes of different function (e.g., esterases and phosphatases) may be structurally much closer to each other than to enzymes of similar function, especially when function is determined by a nonspecific substrate.

Apparently there is no relationship between the variability of an enzyme and its quaternary structure, that is, whether it is monomeric, dimeric, or polymeric. Possible correlations with molecular size have not been sufficiently studied. For phosphoglucomutase and NADP-linked dehydrogenases that supply NADPH for reductive synthesis, it has been suggested that selection against mutants is on the average less intense because they are represented in cells and tissues by several functionally similar enzymes (Cohen *et al.,* 1973). Singh (1975) offers a similar explanation for the greater variability of group II enzymes.

Data for vertebrates summarized by Selander and Johnson (1973) suggest that the mitochondrial isozymes of isocitrate dehydrogenase and glutamic oxaloacetic transaminase are less variable than their corresponding isozymes in the cytosol, but average heterozygosities for the mitochondrial and cytosol forms of malate dehydrogenase are similarly low. An extensive investigation of variability of enzymes in relation to association with membranes is needed.

Our discussion thus far has identified several major sources of heterogeneity in molecular variation. The result is that, notwithstanding the immense amount of effort expended in surveying variation in organisms in the last decade, the sample sizes of loci are generally inadequate for satisfactory analyses of the variation: molecular heterogeneity is too great. This point can be illustrated as follows: Estimates of H for 28 continental forms of *Drosophila* (based on an average of 24 loci) range from 0.04 in *D. busckii* to 0.29 in *D. affinis,* with a mean of 0.1499. The correlation of P with H is 0.764, and the relationship is linear (Figure 4). Because heterozygosity apparently is not uniform over species, it has been necessary to erect explanatory hypotheses, including the suggestion that interspecific differences in variability reflect ecological parameters related to niche width. Some of the observed species differences may be real, but the major determinant of the span of variation in estimates of polymorphism is the laboratory in which the survey was conducted! All surveys have included several esterases and other highly variable group II enzymes. A more significant variable is the proportion of group I enzymes (which have a wide range of heterozygosity) in the sample of proteins examined. When H is plotted against percentage of group I loci surveyed, species of *Drosophila* fall into two almost discrete clusters (Figure 4). Studies from the laboratories of Lewontin and his associates have included less than 40 percent group I loci, and H values fall below 0.14. But in comparable surveys from the laboratories of Ayala and Kojima, which have included 50 percent or more of group I loci, all but one of the values exceeds 0.14.

Lewontin's (1974) estimate that at least 100 loci will be required for adequate determinations of average heterozygosity may be too low. Because available estimates of H are unreliable, it is preferable to consider variation in individual, homologous enzymes when close comparisons of species are made.

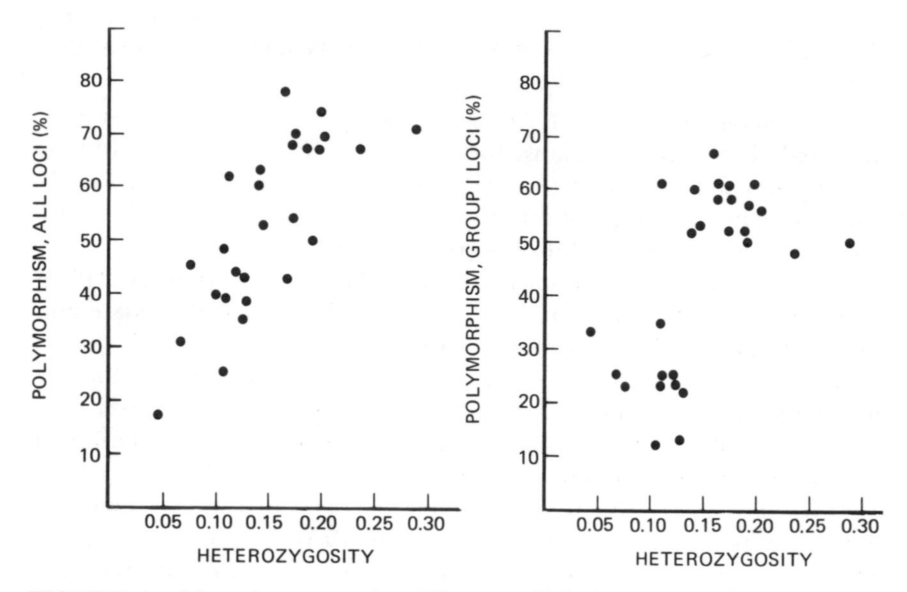

FIGURE 4. Mean heterozygosity (*H*) over all loci examined in relation to polymorphic proportion of all loci (left) and of group I loci (right) for 28 species of *Drosophila*.

VARIATION AMONG MAJOR TAXONOMIC GROUPS

Notwithstanding the statistical limitations of our estimates of variability, it is apparent that not all types of organisms are equally variable (Table V and Figure 5).

Invertebrates. Diploid insects and some groups of marine invertebrates are highly polymorphic. Outbreeding land snails apparently also are highly variable, but estimates for several species of the marine snail genus *Nerita* are low. Variability is low ($H = 0.01$ to 0.06) in large marine crustaceans (lobsters and crabs) (Tracey and Nelson, 1975).

Vertebrates. Estimates of H and P for mammals, reptiles (mostly lizards), and birds are rather uniformly low, being about half as large as comparable values for small invertebrates. Man and other large land mammals seem to be no less variable than rodents. Amphibians apparently are on the average more variable than amniotes. Estimates for fish are highly heterogeneous (in part perhaps because they have come from a large number of laboratories). Recent studies by Gary Sharp (personal communication) suggest that levels of variability are unusually low in large marine vertebrates such as tuna fish and porpoises.

Plants. Few plants species have been studied, and estimates are highly heterogeneous. However, the data suggest that polymorphic levels are high in many species (Levin, 1975; also see Chapter 8).

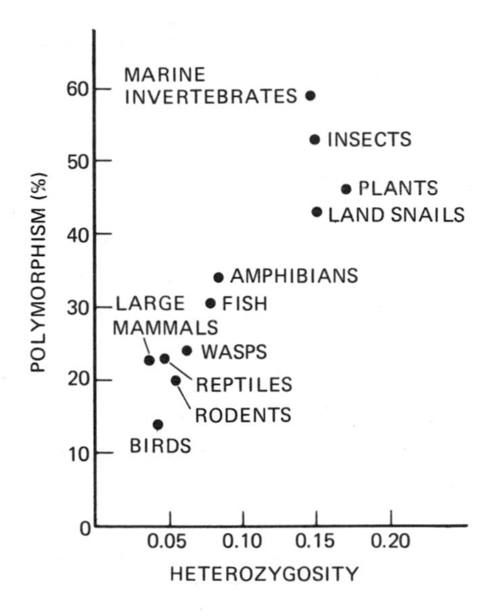

FIGURE 5. Genic variation in some major groups of animals and in plants. Mean values of *P* and *H* are plotted. (See Table V.)

TABLE V. Genic variation in some major groups of animals and in plants

Group	Number of species or forms	Mean number of loci per species	Mean proportion of loci: Polymorphic per population	Mean proportion of loci: Heterozygous per individual
Insects				
Drosophila	28	24	0.529 ± 0.030	0.150 ± 0.010
Others	4	18	0.531	0.151
Haplodiploid wasps	6	15	0.243 ± 0.039	0.062 ± 0.007
Marine invertebrates	9	26	0.587 ± 0.084	0.147 ± 0.019
Snails				
Land	5	18	0.437	0.150
Marine	5	17	0.175	0.083
Fish	14	21	0.306 ± 0.047	0.078 ± 0.012
Amphibians	11	22	0.336 ± 0.034	0.082 ± 0.008
Reptiles	9	21	0.231 ± 0.032	0.047 ± 0.008
Birds	4	19	0.145	0.042
Rodents	26	26	0.202 ± 0.015	0.054 ± 0.005
Large mammals*	4	40	0.233	0.037
Plants†	8	8	0.464 ± 0.064	0.170 ± 0.031

*Man, chimpanzee, pigtailed macaque, and southern elephant seal.
†Predominantly outcrossing species; mean gene diversity is 0.233 ± 0.029.

HYPOTHESES TO ACCOUNT FOR PROTEIN POLYMORPHISMS

Two major hypotheses have been advanced to explain variation in average level of heterozygosity among taxonomic groups of organisms. The neutralists' interpretation (Kimura and Ohta, 1971*b*; Nei, 1975) is that it reflects differences in population size (number of individuals in the species), the more variable organisms having larger populations. (See discussion by Soulé, Chapter 4.) But this hypothesis is difficult to test because we cannot determine the evolutionarily significant effective size of populations, which corresponds roughly to the harmonic mean number of breeding individuals over generations (Nei *et al.,* 1975).

Selander and Kaufman (1973*a*) attempted to account for variation in level of genic heterozygosity in terms of Levins' (1968) theory of adaptive strategies in relation to environmental uncertainty, or *grain.* According to this hypothesis, organisms with generally larger body size, greater mobility, and greater physiological and behavioral homeostatic control tend to experience the environment as fine-grained, and, hence, the optimum strategy is more often a single phenotype specialized to the most frequently encountered set of conditions. Smaller, less mobile organisms with relatively poor homeostatic control experience the environment as coarse-grained, and for them the optimum strategy is more often the development of specialized morphs occurring in proportions dependent upon the frequencies of the different patches of environment. Using a different approach, Gillespie (1974) and Gillespie and Langley (1974) also concluded that genic variation should diminish with increasing homeostasis.

In advancing the homeostasis theory, Selander and Kaufman (1973*a*) rejected the species-number hypothesis largely on the basis of preliminary evidence (see Johnson *et al.,* 1975) that the endemic Hawaiian species of *Drosophila* are on the average as polymorphic as the continental species, although many of them are confined to single islands and have numbers several orders of magnitude smaller than continental species. Ayala's (1975, and personal communication) recent analysis of 60 Hawaiian species has now confirmed earlier reports; H values range from about 0.10 to 0.20, with a mean of about 0.15. The homeostasis theory also is supported by evidence of reduced levels of variability in large, mobile marine crustaceans and vertebrates, all of which have very large populations.

It has been suggested (Nevo and Shaw, 1972; Nevo *et al.,* 1974) that subterranean and fossorial rodents (pocket gophers and the mole rat) have unusually low levels of heterozygosity as a consequence of selectively mediated responses to a "monotonous subterranean niche." However, the available data fail to support this thesis (Table VI). Mean heterozygosity for strongly fossorial species is not significantly different from

TABLE VI. Genic heterozygosity in rodents

Genus	Number of species	Heterozygosity* Mean	Range
Fossorial			
Spalax	1	0.034	
Thomomys	3	0.049	0.03–0.07
Geomys	3	0.061	0.05–0.07
Mean (7 species)		0.0520 ± 0.0059	
Other			
Dipodomys	5	0.033	0.01–0.07
Peromyscus	8	0.061	0.03–0.10
Mus	1	0.084	
Sigmodon	2	0.036	0.03–0.04
Eutamias	1	0.063	
Mean (17 species)		0.0516 ± 0.0066	
Mean (12 species)†		0.0592 ± 0.0066	

*To compensate for an absence of esterases among the proteins sampled, reported *H* values for several species were increased by 43 percent, the mean contribution of esterases to overall heterozygosity in 11 species of rodents.

†Excluding *Dipodomys.*

that for nonfossorial species. Although heterozygosity tends to be somewhat low in the subterranean (burrowing) kangaroo rats (*Dipodomys*), it is "normal" or high in other burrowing species, including *Peromyscus polionotus* (0.060) and *Mus musculus* (0.084). Similarly, Somero and Soulé (1974) found that marine fish from thermally variable habitats are not more polymorphic than species from habitats with stable thermal regimes, and data gathered by Gooch and Schopf (1973) and Ayala and Valentine (1974) indicate that the fauna of the deep sea, a supposedly stable and homogeneous habitat, is not genetically impoverished. Various aspects of genic variability in relation to environmental heterogeneity are discussed at length in Chapter 5.

REDUCTION OR LOSS OF VARIABILITY IN SMALL POPULATIONS

In organisms inhabiting isolated islands or caves where populations were founded by a few individuals or are very small and subject to periodic bottlenecking, genic variability may be severely reduced or even absent. This phenomenon has been reported for rodents (Selander *et al.*, 1971; Hunt and Selander, 1973; Selander *et al.*, 1975), lizards (Webster *et al.*, 1972; Gorman *et al.*, 1975), seals (Bonnell and Selander, 1974), fish (Avise and Selander, 1972), and insects (Prakash, 1972; Saura *et al.*, 1973).

The founder effect and intermittent random drift probably are the major causes of reduced variability in isolated populations, but fixation of alleles has also been attributed to directional selection (Soulé, 1973; Gorman *et al.,* 1975).

Genetic drift has also been invoked to explain low genic variability ($H = 0.017$) within and marked differentiation between troops of the Japanese macaque (*Macaca fuscata*), whose effective population size is estimated at 20 individuals (Nozawa *et al.,* 1975).

Extensive work on *Drosophila* has failed to demonstrate reduced genic variability in populations that are geographically or ecologically marginal and frequently are depauperate in chromosomal inversion polymorphism (Prakash *et al.,* 1969; Ayala *et al.,* 1971; Prakash, 1973; Zouros *et al.,* 1974; discussion in Soulé, 1973).

BREEDING SYSTEMS AND VARIABILITY

Haplodiploidy

Estimates of H and P for solitary and social wasps (Lester, 1975) and for several species of bees (Snyder, 1974) indicate that variability is severely reduced in the haplodiploid Hymenoptera. As shown in Table IV, wasps are only 38 percent as heterozygous as *Drosophila* and 26 percent as heterozygous as other (diploid) insects. The reduction is fairly consistent over loci, with $r = 0.832$ ($p < 0.01$) for the comparison of 12 loci in wasps and *Drosophila.*

Although the low variability of haplodiploid insects might seem highly relevant to the controversy regarding the causal basis of protein polymorphism, interpretation is in fact extremely difficult because several different theories have predicted it (Lester, 1975). When one sex is haploid, polymorphism is less likely to be maintained by balancing selection (Hartl, 1971), and equilibrium frequencies in the mutation-equilibrium model of Ohta (1974; see discussion beyond) will generally be lower. Moreover, effective population size may be a factor because the species of wasps and bees that have been studied are not abundant.

Parthenogenesis

Parthenogenetic species that have arisen through hybridization of two or more bisexual species understandably may show very high levels of heterozygosity. The heterozygosity of the original hybrid persists at many or all loci, as a consequence of the absence or rarity of recombination. In diploid and triploid populations of the parthenogenetic lizard *Cnemidophorus tesselatus,* average heterozygosity is 0.560 and 0.714, respectively, as compared with 0.04 to 0.08 for the bisexual species of the genus (E. Davis Parker, personal communication). Mean heterozygosity is 0.314 in triploid and 0.390 in tetraploid populations of the parthenogenetic weevil *Otiorrhynchus scaber* (Suomalainen and Saura, 1973). (Heterozygosity in a diploid bisexual population of *O. scaber* studied by Suomalainen and Saura, 1973, also was unusually high, 0.304 over 24 loci, whereas estimates for three populations of the related diploid bisexual species *Strophosomus capitatus* were within the range reported for *Drosophila* and other insects.) For tetraploid populations of the moth *Solenobia triquetrella,* Lokki *et al.* (1975) estimated mean heterozygosity at 0.203.

Self-fertilizing species

Studies of allozymic variation in plants (Allard and Kahler, 1971) and snails (Selander and Kaufman, 1973*b*) have fully confirmed other lines of evidence (Allard *et al.*, 1968) indicating that strongly self-fertilizing species maintain levels of genetic variation at least equal to those of outbreeding forms. However, because of the effects of close inbreeding in reducing heterozygosity, the variation is distributed largely among strains (families).

Neither Allard *et al.* (1975) nor Levin (1975) could find a clear relationship between the mating system and amount of variability in plants. For example, the slender wild oat *Avena barbata* outcrosses much more frequently but is much less variable genetically than *A. fatua.*

In populations of the self-fertilizing land snail *Rumina,* which occurs in southern France as two ecologically differentiated strains, there is no heterozygosity apart from that generated locally by outcrossing between strains (Selander and Hudson, 1975). However, genic diversity in *Rumina* is as great as that of outcrossing forms such as *Helix* and *Cepaea,* because different alleles are fixed at 50 percent of the structural gene loci of the strains.

Clonal variation in *Escherichia coli*

Milkman (1975) has studied variation in proteins encoded by five loci in 829 clones of *E. coli* from 156 fecal samples of diverse origin and examined his findings with reference to models of the relationship between neutral allelic variation and electromorphic (mobility) variation. If one assumes that mobility variants are limited to integer classes (Ohta and Kimura, 1973; Nei and Chakraborty, 1973; King, 1973), the steady-state relationship, which is reached after $4N_e$ generations at population size N_e, is

$$m_e = (1 + 8N_e v)^{\frac{1}{2}}$$

where m_e is the effective number of electromorphs and v is the rate of neutral or nearly neutral mutation. If $N_e = 10^{10}$ (a small fraction of all *E. coli* in the world) and $v = 10^{-8}$ (presumably a minimal estimate), the expected effective electromorph number is between 25 and 30; but in fact Milkman observed only one or two in each of five cases. (The actual number of electromorphs varied from three for alkaline phosphatase to 12 for 6-phosphogluconate dehydrogenase, but for each protein one electromorph is in high frequency in all populations.) If the electrophoretic method employed by Milkman is sufficiently sensitive to resolve most existing electromorphic variation in the molecules studied, his findings clearly are incompatible with the neutral theory. However, they are not contrary to predictions of the mutation-equilibrium model, which postulates convergent generation of a relatively small number of allelically heterogeneous electromorphs.

RECENT DEVELOPMENTS IN THEORY

It is now apparent that the neutralist theory, as it was originally developed brilliantly by Kimura (1968, 1969) and King and Jukes (1969), does not adequately explain protein polymorphism and evolution. However, if this theory cannot account for all aspects of variation in natural populations and is incompatible with certain experimental evidence from the laboratories of molecular biologists and population geneticists, it does not follow that the case for balancing selection and adaptation has been made. Exploring the consequences of very slightly deleterious (but tolerable) mutations, with selection coefficients on the order of the mutation rate, the former neutralists have produced a second-generation non-Darwinian theory of molecular polymorphism and evolution, which can be called the mutation-equilibrium theory. Following Ohta (1974), the main elements of the new theory may be summarized as follows.

There is a continuum of mutations from neutral ones, which have neither advantageous nor detrimental phenotypic effects, to those mutations on which selection can act. The probability that a mutant will reach a certain intermediate frequency or will be fixed in a population depends on its selection coefficient. If selectively neutral (i.e., equivalent phenotypically to the wild-type allele), the rate of substitution is equal to the mutation rate (Kimura, 1969). If advantageous, the substitution rate is very much higher; and if detrimental, lower.

Because the maximum observed evolutionary rate of proteins is close to estimates of the intrinsic mutation rate, it is assumed that only a few amino acid sites of a protein molecule are replaced in any one period of evolutionary time by definitely advantageous mutations. Molecules evolve at rates determined by the average selection coefficients of new mutations, which in turn are dependent upon the structural and functional constraints of the molecules. Once the structure and function of a molecule are determined in the course of evolution, natural selection acts mainly to maintain them (stabilizing selection) and then becomes, in a sense, mostly negative. Variant alleles of slightly decreased fitness reach frequencies determined by the equilibrium between mutation and selection. Thus, positive Darwinian selection is assigned a minor role in molecular evolution.

The neutral theory cannot adequately account for the geographic uniformity of molecular polymorphism, combined with relatively small numbers of alleles, in species of *Drosophila* and some other organisms. If migration sufficient to account for the uniformity is invoked (Kimura and Maruyama, 1971; Maruyama and Kimura, 1974), effective population size is inflated, and large numbers of alleles are expected (Ayala, 1972). This difficulty disappears if, as now suggested, electromorphs are allelically heterogeneous charge classes that can convergently evolve in populations. Several models of stepwise integer production of electromorphs have been investigated (Ohta and Kimura, 1973; King, 1974; Ohta and Kimura, 1975; King and Ohta, 1975). In these models, which were suggested by the observation that the less frequent electromorphs tend to cluster symmetrically about the common ones (see analysis by Richardson *et al.*, 1975), mean selection coefficients may be equal over electromorphs or may increase with increasing distance of an electromorph from that of the "type allele." In any event, it is proposed that polymorphism can be maintained in large populations by a balance between mutation to slightly deleterious alleles (whose products are manifested as electromorphs at integer states on gels) and selection against the mutants. Provisions of the models are that the product of the effective population size and the mutation rate is large and that the selection coefficient is larger than the mutation rate. The distribution of electromorphs is determined by the ratio of the selection coefficient to the mutation rate.

One advantage of the mutation-equilibrium model over the neutral theory is that an upper limit to heterozygosity in large populations can be specified. In small populations, very slightly deleterious alleles become effectively neutral, and heterozygosity is determined by the product of the effective population size and the mutation rate, as in the neutral model (Ohta and Kimura, 1974). Because the average selection coefficient for mutants is smaller in small populations, variability at equilibrium may be as high as in larger populations, and unusually low levels of variability in populations or in species can be explained by postulating that they have not yet reached mutation-selection equilibrium (Ohta, 1974). Because the rate of evolution is expected to be higher in small populations, such as occur at the time of speciation, the new theory can account for the category of loci with nonoverlapping allele distributions between sibling species. (See Ayala and Gilpin, 1974.) And the occurrence of the same set of electromorphs in similar frequencies in different species can be attributed to convergent mutation-selection equilibria following bottlenecking at the time of speciation.

Because average selection coefficients of mutants will be determined by the structural and functional constraints of molecules, variation in level of heterozygosity among different types of enzymes also can be accounted for by the new theory.

It should be obvious that the mutation-equilibrium theory is very powerful and makes many of the same predictions as does the selectionists' model. One distinguishing feature of the new theory is that it predicts extensive heterogeneity of electromorphs within and, more importantly, among populations. Therefore, the determination of the true allelic composition of electromorphs has become an important aspect of research in molecular population genetics.

Evidence of heterogeneity in electromorphs is beginning to accumulate from studies of *Drosophila* and other organisms, but as yet we have no precise measurements of its extent over either loci or species. For example, Bernstein *et al.* (1973) reported that each electromorph of xanthine dehydrogenase in several species of the *D. virilis* group consists of an average of 2.9 polypeptides that are distinguishable by heat-stability characteristics. Because these authors found that the heat-sensitive alleles are strongly differentiated geographically in species, whereas electromorphs are not, they suggested that the latter are maintained by balancing selection while the former are selectively neutral and, hence, drifting in local populations. However, this interpretation was rejected by Ohta (1974; see Ohta and Kimura, 1975) in favor of the view that the electromorphs are in mutation-selection balance in the species as a whole, whereas the individual component alleles are not because of lower mutational input.

Because it invokes the deterministic factor of selection, the mutation-equilibrium theory might be viewed as a compromise with the selectionists' theory. This would be erroneous. The new theory, which is merely an extension of the neutral hypothesis, is directly opposed to the position that molecular variation is maintained in populations by heterotic, frequency-dependent, or other types of balancing selection and that directional selection is the major cause of allele fixation. The neutralists have yielded no real ground, and the controversy will continue.

PROSPECTS IN MOLECULAR POPULATION GENETICS

The appearance of this volume is appropriately timed to summarize the first phase of the molecular study of biological evolution, which for population genetics began a decade ago with the discovery of tremendous amounts of protein sequence variability within and among species of organisms. It was soon realized that the basis for maintenance of protein polymorphism in populations must be explained if we are to understand the mechanisms of evolution at the molecular level (Kimura and Ohta, 1971*a*). However, although efforts to characterize this new and unfamiliar type of variation have stimulated major changes in theoretical population genetics and have contributed to an understanding of the genetics of speciation and mating systems, we have made little progress in defining the processes of molecular evolution. The problem is that geneticists have had great difficulty determining what part of variation is functionally relevant, that is, the "stuff of evolution." As Lewontin (1974) has noted, in using molecular data to test the "classical" and the "balanced" theories of the genetic structure of populations we may have been asking the wrong question: "That is, the question was never really, How much genetic variation is there between individuals? but rather, What is the nature of genetic variation for *fitness* in a population?" A heated and complex controversy has developed between those who propose that the random fixation of neutral or nearly neutral mutants is the main cause of molecular evolution and those who believe selection to be the important factor.

The study of molecular population genetics has entered a new phase in which hypothesis testing will be even more difficult because rival theories make less distinguishable sets of predictions. Success or failure of our attempts to understand molecular evolution will depend on our ability to identify the genetic unit of selection, to obtain data relevant to it, and to develop appropriate statistical methods of analysis (Lewontin, 1974; Sing and Templeton, 1975). If fitness differentials of genotypes at most loci are too small to be detected in laboratory experiments, we may never acquire a clear understanding of some important processes of molecular evolution.

Electromorphs may prove to be a very special class of variants, as suggested by Markert (1968) and Kimura (1971). We first pretended that they were not, in order to justify our attempts to measure genic variation by the zymogram technique; but if electrostatic charge is an important adaptive character of proteins, electromorphs may be rare compared with variants not affecting the charge of polypeptides. If so, our present estimates of genic variation in populations are heavily biased and will not be useful in solving some of the more important theoretical problems now being formulated. Attempts now underway to assess allelic heterogeneity of electromorphs by indirect biochemical methods will at least provide an indication of the scope of the problem. Thus far we have been able to proceed with minimal formal genetic work because electromorphic variation is relatively easy to interpret; but as we begin to explore other levels of diversity, the identification of allelic protein variants by heat denaturation, thiol reagents, and other methods must be verified by thorough genetic analysis. This means that only *Drosophila, Mus,* and a few other organisms are suitable material for these studies.

Significant advances in our understanding of the adaptive basis for protein polymorphism will come from long-term efforts involving population geneticists and biochemists. Recent work on the polymorphic alcohol dehydrogenase locus in *D. melanogaster* (Clarke, 1975) is a model of this approach.

Electrophoresis will be increasingly employed as a powerful tool in systematics (Avise, 1974; and Chapter 7) and a means of demonstrating markers useful in the analysis of population structure (Crow and Denniston, 1974; Selander, 1975). It is hoped that it will not prove to be a dead end as far as the major problems of evolutionary genetics are concerned, although this remains a real possibility (Sarich, 1972*b;* King, 1975).

In his excellent book on evolution and the new biology, Nigel Calder (1973) concludes that "To leave the New Darwinism of 1960 intact only a complete rout of Kimura and his fellow heretics will do. Such an outcome seems less likely with every year that passes. . . . Whatever the outcome, molecular biology has already made an indelible mark on studies of evolution." Ten years ago evolutionary biology was much less active and intellectually rewarding than today, and its historical position as a discipline fundamental to all of biology seemed threatened by the continuing rise and increasing dominance of molecular biology. Evolutionists were becoming apologists, denying that their field was moribund. In the last decade, however, evolutionary biology has been revitalized and enriched through contact with molecular biology and largely as a consequence of efforts to characterize protein variation and to solve the difficult and intriguing problems it presents for evolutionary theory. Considering the magnitude of this effect, we may not be overfanciful to think that future historians will see molecular biology more as the salvation for than, as it first seemed, the nemesis of evolutionary biology.

SUGGESTED READINGS

The outstanding review of electrophoretic studies of protein variation from the perspective of population and evolutionary genetics is R. C. Lewontin, *The Genetic Basis of Evolutionary Change,* Columbia University Press, New York, 1974.

The neutrality theory of protein polymorphisms is presented by M.

Kimura and T. Ohta, Protein polymorphism as a phase of molecular evolution, *Nature,* **229,** 467–469, 1971. For the recent mutation-equilibrium theory see J. L. King and T. H. Ohta, Polyallelic mutational equilibria, *Genetics,* **79,** 681–691, 1975. Empirical evidence against the neutrality theory is advanced in F. J. Ayala, Biological evolution: natural selection or random walk?, *Amer. Sci.,* **62,** 692–701, 1974; and in M. T. Clegg, R. W. Allard, and A. L. Kahler, Is the gene the unit of selection? Evidence from two experimental plant populations, *Proc. Nat. Acad. Sci. U.S.A,* **69,** 2474–2478, 1972.

The proceedings of a symposium dedicated to the selection versus neutrality controversy deserve consultation: L. M. Le Cam, J. Neyman, and E. L. Scott (eds.), *Proc. Sixth Berkeley Symp. Math. Stat. Prob.,* vol. V, University of California, Berkeley, 1972.

CHAPTER 3

GENETIC POLYMORPHISM AND ENZYME FUNCTION

GEORGE B. JOHNSON

A great deal of information has been gathered on the three-dimensional structures of proteins since the structures of the heme proteins first became known through X-ray crystallography almost two decades ago. Detailed information is now available on over 15 proteins (Chothia, 1975). A general picture has emerged of proteins as closely packed structures. The noncovalent forces which maintain these folded structures result from lowered entropy because of reduction of nonpolar (hydrophobic) contacts with water (Kauzmann, 1959) and because nearly all buried polar groups form hydrogen bonds. This interior hydrogen bond formation imposes very severe geometrical requirements (Chothia, 1975) and limits the range of possible stable tertiary (three-dimensional) structures which a given protein can assume.

The most biologically significant fact about our emerging view of protein structure is that each individual bond involved in forming and stabilizing protein structure is weak, with energies no more than an order of magnitude greater than the thermal energies of the solvent molecules. The cumulative effect of these weak bonds is substantial, however. The *collective* interaction of amino acid side groups with each other and with the solvent (water) is the major force that maintains the tertiary structure. Minor changes in either side group or solvent can have major effects (Mauridis *et al.,* 1974). This sensitivity provides the basis of a particularly important general property of proteins, *allosterism.* The word "allosteric" means "other form" and refers to the fact that a protein's three-dimensional shape is often altered when ions or small molecules bind to it. The binding energy contributed by the small molecule, if sufficiently high, alters the equilibrium of forces maintaining the protein's shape (conformation); then one of a limited number of alternative conformations is assumed. A protein's susceptibility to such binding can be made highly specific when binding sites are specific for particular small molecules. When the protein is an enzyme, a change in its shape can change its catalytic properties. Thus, the property of being structurally responsive to low-molecular-weight *effector* molecules has permitted the evolution of metabolic control systems of great sensitivity, in which metabolites are used as feedback signals to modulate the activity of enzymes. Most of intermediary metabolism is coordinated by such controls. Levels of ATP, for instance, modulate the activity of many related pathways which produce and utilize energy (Atkinson, 1971). Thus allosteric modulation of protein tertiary structure is the basic tool employed in the integration and coordination of metabolism. In other words, the flexibility of protein structure permits small molecules to act as signals in the cell to regulate the cell's activities.

One can imagine that an allosteric enzyme ought to be quite sensitive to variation in cellular effector concentrations or to changes in temperature. Take as an example an enzyme adapted to operate at low temperature: The reversible conformational changes required for allosteric control at low temperature imply a nonrigid tertiary structure. Low temperatures provide the system with less kinetic energy, and only a flexible structure will be capable of undergoing a conformational transition when a relatively small amount of energy is added to the protein by the binding of the allosteric effector molecule. On the other hand, an enzyme adapted to operate at a higher temperature needs to be far more rigid if it wishes to remain active and to carry out reversible conformational changes in the presence of far more thermal kinetic energy. Thus, it follows from the fundamental nature of protein structure that a given enzyme is capable of allosteric modulation only over a restricted range of conditions. An enzyme operating at low temperatures must be relatively flexible, while an enzyme operating at high temperatures must be more rigid. Adaptation to markedly different conditions would require changes in the amino acid sequence of the protein.

From this we would predict that evolutionary adaptation to different habitats must entail significant alteration of amino acid sequence, particularly in the case of allosterically modulated regulatory enzymes. We have known for a decade that large amounts of protein genetic variability exist in nature (Lewontin and Hubby, 1966; Harris, 1966). The question thus arises as to whether this variability, detected by electrophoretic screening, reflects the differential adaptation which the above physiological considerations have led us to expect.

ENVIRONMENTAL AND GEOGRAPHIC CORRELATIONS

If electrophoretically detected protein variants indeed reflect differential adaptation, then one might predict that the amount of such variability would correlate with environmental diversity—that is, that the number of available niches would in some sense dictate the level of variability. A variety of studies have addressed this issue. A typical result is that reported by Somero and Soulé (1974), who have estimated levels of variability in tropic, temperate, and Arctic species of fish. They found the fish species of temperate regions to be far less polymorphic than tropic species. Thus, in this study protein variability is seen to correlate with regions of low thermal diversity. A variety of other studies report a correlation between levels of protein variability and latitude, species diversity, etc. (Johnson, 1974*b;* and Chapter 5). These results have proved difficult to interpret, however. It is not clear whether variability in the tropics should be seen as reflecting stability of some environmental factor, such as temperature, or variation of some factor, such as diversity of available food (which may be true in the tropics because of high species diversity). The difficulty arises from the experimental question being too broadly phrased. For a successful approach to this question, an examination of levels of variability at specific loci in habitats differing in ways known to affect the functioning of the enzymes specified by those loci is required. Unless the selectively important parameters are specifically identified, one cannot state with any precision what pattern of variation differential adaptation would be expected to produce.

Thus, generalized studies of macrogeographic patterns of protein variability have suggested that more clearly defined empirical systems are needed.

Considerable data are available on specific enzyme systems, but care needs to be exercised to avoid spurious conclusions. An example is the enzyme alcohol dehydrogenase (ADH) as studied in *Drosophila melanogaster*: A marked latitudinal cline is reported along the United States eastern seaboard, with Florida essentially monomorphic and New York highly heterozygous (Vigue and Johnson, 1973). Allele frequencies are shown to correlate with a variety of characteristics of the habitat. It has since been learned, however, that ADH in this species is closely linked to a small inversion (Mukai, personal communication) which exhibits the same latitudinal cline in frequency. Thus, it is not clear whether the target of presumptive differential selection is the ADH locus or the linked inversion. Similar confusion about the selective role of the environment with respect to this locus arises from studies which assume that the function of the ADH enzyme detected by electrophoretic screening is to catabolize excessive levels of ethanol (Gibson, 1970). It is now clear that in *Drosophila* a variety of other physiological processes are catalyzed by alcohol dehydrogenases, among them retinene to retinol, a reaction critical to vision, and glyceraldehyde to glycerol, a key entrance into fatty acid metabolism. The selective forces acting on each of these ADH enzyme loci must be very different. Only when studies of ADH can document clearly the specific physiological function of the enzyme actually under study can environmental patterns of ADH variation be interpreted with confidence.

A few loci have been studied in which the physiological function seems clear. These loci are responsible for maintaining NAD/NADH oxidation-reduction potentials (redox levels) in the cell [lactate dehydrogenase (LDH) for vertebrates, α-glycerophosphate dehydrogenase (α-GPD) for invertebrates]. Redox modulating enzymes, along with "energy-charge" (ATP/AMP) modulating enzymes and fatty acid desaturation enzymes, provide a means whereby a cell can coordinate its metabolic response to a generalized physiological perturbation such as is caused by a change in temperature. Where examined, these loci do seem to exhibit variability which correlates with temperature: (1) At the LDH locus of freshwater fish, the water temperature appears to dictate which allele is present, exactly as an adaptive hypothesis would predict (Merritt, 1972). (2) The α-GPD loci of *Drosophila* are generally far more polymorphic in temperate than in tropic populations (Johnson, 1974*b*). (3) In *Colias* butterflies, only populations living in the thermally variable montane habitat are polymorphic for α-GPD, while populations of Alpine and lowland areas (whose temperatures do not vary so much) are not (Johnson, 1972). Furthermore, when a single population of *Colias* straddles timberline, occupying both Alpine and montane habitats, a marked cline in α-GPD heterozygosity is seen within the single population (Johnson, 1975)!

Thus, in a few cases variation is seen to correlate with environment as an adaptive hypothesis would predict. This does not establish a cause-effect relationship, however. Alternative hypotheses, such as linkage to another locus under selection, are not excluded. In an attempt to investigate more directly possible cause and effect relationships, a variety of studies have been carried out involving artificially imposed selection acting upon laboratory populations. The results have been mixed. When a variety of loci of *Drosophila melanogaster* are examined and the habitat varied in several ways, protein variability is seen to correlate with habitat variability (Powell, 1971; McDonald and Ayala, 1974); while when a single locus, esterase-5, is monitored, no such correlation is seen (Yamazaki, 1971; see, however, Marinković and Ayala, 1975). The confusion results from the same ambiguity which confused the studies of macrogeographic patterns in natural populations: in no case are the habitat factors which are varied of demonstrable importance to the locus under study. Thus, for lack of an empirically sufficient experimental approach the experimental evaluation of putative environmental and geographic correlations with protein variation remains ambiguous. What the experiment requires is a known change in *selection,* not simply a change in environment.

THE NATURE OF BALANCING SELECTION

Despite the ambiguity of the studies discussed above, most investigators have concluded that the observed protein variation *is* adaptive and that it is maintained by some form of balancing selection. Two basic suggestions have emerged to account for what would have to be very widespread balancing selection. One has its roots in the thinking of Levene, Beardmore, and Levins (Levene, 1953; Beardmore and Levine, 1963; Levins, 1968) and proposes that protein polymorphism reflects a variable niche and represents an evolutionary strategy for dealing with a heterogeneous environment. The emphasis is on environmental disruptive influences, which serve to maintain the genetic polymorphism in a dynamic equilibrium. The alternative suggestion derives from the thinking of molecular biologists (Fincham, 1972) and proposes that protein polymorphism derives from the inherent superiority of the hybrid *heterozygote* molecule. The emphasis is on inbuilt heterosis which serves to maintain the polymorphism despite a heterogeneous environment. These two hypotheses are fundamentally different in their view of the environment and serve to distinguish two very different viewpoints concerning the action of natural selection. Like the paradigms described by Lewontin (1974), the first hypothesis involves a dynamic viewpoint, the second a static one.

THE HYPOTHESIS OF BALANCED ADVANTAGE

The hypothesis of balanced advantage reflects a classic view of polymorphism as a dynamic equilibrium resulting from contrasting ongoing disadvantages (Ford, 1965). The concept is based ultimately upon the existence of counterbalancing selectively important variation in the habitat. A similar view may be taken of protein polymorphism as maintained by a dynamic balance of environmental heterogeneity. A well-known example of such a molecular polymorphism maintained by offsetting disadvantages is the case of sickle cell hemoglobin (Ingram, 1957). While few such cases have been documented, there are many indications that allelic alternatives at enzyme loci produce proteins that have the potential to interact differentially with the environment and there exist some indirect inferences that this potential is realized.

That alternative allozymes are capable of responding differentially to changes in their immediate environment is essential to any concept of allozymes as functionally distinct. A wide variety of investigations of isozyme and allozyme polymorphisms over the last 10 years have uniformly reported significant differences in kinetic behavior. Differences between variants are seen in binding affinity of enzyme for substrate or cofactor, in the thermal sensitivity of such binding, in the maximal reaction rate, in stability to high temperature, etc. There seems little question in the cases which have been investigated that electrophoretically detected enzyme variants differ significantly in their kinetic behavior.

Another line of very strong but indirect evidence that allozymes of polymorphic loci are functionally different comes from the observation that they have quite different three-dimensional shapes (conformations). Among 15 loci of a butterfly species, over 90 percent of the electrophoretically different proteins are conformationally distinct, as judged by differential sieving behavior on polyacrylamide gels, although in many cases no charge change is involved (Johnson, 1975). It is difficult to believe that such generalized conformational differentiation is not reflected in functional differences.

Another indirect argument for functional differentiation is provided by comparing variation at loci of enzymes in different functional classes. When levels of polymorphism in metabolically different classes of enzymes are compared, the results uniformly suggest evironmental involvement. Loci whose enzymes utilize substrates originating from the external environment are far more polymorphic than loci whose enzymes utilize internal metabolites (Gillespie and Kojima, 1968; Kojima *et al.*, 1970; Johnson, 1973) (Table I). This suggests that genetic polymorphism reflects a physiological response to environmental variation.

In general, then, there is little direct evidence (other than the classic case of sickle cell hemoglobin) that protein polymorphism is maintained by a dynamic balance of contrasting environmental forces. Available evidence indicates that the *potential* for functional differentiation between allozymes exists and suggests environmental involvement, but it has not demonstrated that environmental heterogeneity itself is the vehicle of selection.

THE HYPOTHESIS OF HETEROLOGOUS ADVANTAGE

The second hypothesis discussed above is more generally accepted. This hypothesis has been conditioned by analogy to enzyme complementation (Fincham, 1972). Many enzymes comprise separate identical subunits. Such enzymes are called multimeric, as opposed to single-subunit

TABLE I. Allozyme diversity in *Drosophila* correlates with variation in substrate. (After Johnson, 1973)

Drosophila Species	Enzymes using substrates from within the cell		Enzymes using substrates from the external environment	
	S	*k*	*S*	*k*
subobscura	3	2.67	7	3.14
equinoxialis	2	2.00	4	2.50
pseudoobscura	3	2.67	5	3.60
obscura	3	2.00	6	3.60
willistoni	11	2.84	7	3.55
melanogaster	10	1.48	5	2.86
paulistorum	3	1.31	7	2.86
bipectinata	11	2.27	6	3.67
parabipectinata	11	1.54	6	2.50
malerkotliana pallens	11	1.82	6	3.67
athabasca	9	1.33	5	3.40
simulans	10	1.40	5	3.60
affinis	7	1.43	5	4.00
tropicalis	2	1.00	4	3.25

S: number of loci with samples of at least 100 haploid genomes. *k*: mean number of alleles observed at a frequency equal to or greater than 0.01 in samples with 100 or more genomes. The external environment is considered as being more variable than the interior of the cell.

enzymes, which are dubbed monomers. In a heterozygous cell it is possible to form a heterologous multimer, or hybrid enzyme, with some subunits deriving from each allele. If the heterologous multimer is functionally superior to either homologous nonhybrid multimer, then it is easy to see how polymorphism would be produced by balancing selection. Selection would act to optimize the frequency of the heterologous form, producing a direct heterosis. The result would be that both alleles would be selected to remain in appreciable frequency.

This simple model of heterosis due to superiority of the heterologous protein has been a very popular one. It permits straightforward theoretical treatments, because selection is viewed as acting directly on allele differences *per se*. The heterologous molecule is often viewed as inherently superior, being selectively advantageous in any environment. This contrasts sharply to the preceding view of protein polymorphism as directly reflecting a heterogeneous environment and resulting from a dynamic balance of opposing selective influences. Much of the theory of population genetics implicitly assumes a model of heterologous advantage.

It is a model subject to direct test. Hybrid enzymes may be formed *in vitro* and their functional properties compared with the homologous nonhybrid forms. Such a direct test of overdominance (heterosis) at the biochemical level has recently been reported for the highly polymorphic esterase-5 locus of *Drosophila pseudoobscura,* with little evidence of heterologous advantage (Berger, 1974). In general the hybrid multimers were no better than their homologous counterparts, and in several cases they were worse. This work thus provides direct evidence against the hypothesis of heterologous advantage.

Thus, the available molecular evidence argues rather pointedly against the hypothesis of heterologous advantage as the basis of most heterosis. On the face of it the hypothesis had always seemed biochemically unrealistic: How was it to explain widespread polymorphism at monomeric enzyme loci?

MULTILOCUS APPROACHES

Neither of the two most prevalent views, environmental balance and heterologous advantage, has provided a very satisfactory framework for evaluating the adaptive role of enzyme polymorphism. The problem has not been with the hypotheses so much as with their underlying assumptions. These hypotheses are basically single-locus conceptualizations that view each allozyme independently of other enzymes. This narrow view ignores one of the most important lessons of modern biochemistry—metabolism is a *coordinated* process. The many reactions of intermediary metabolism are subject to interlocking controls which modulate and balance the flux through the various pathways. The Krebs cycle does not proceed independent of glycolysis, and both affect pentose and fatty acid metabolism. Coordinate control of metabolism represents an evolutionary achievement of the first magnitude and must be taken into account when one considers the possible functional effects of an allelic substitution.

A realization that metabolic coordination may be of great evolutionary importance leads one inextricably to multilocus approaches. Existing data on coordination of genotypes are scanty but suggestive. Loci of five hydrolytic enzymes in *Avena* show evidence of coordinate organization (Allard and Kahler, 1973). In *Colias* butterflies I have seen a variety of alternative microgeographic patterns in allele frequency at 12 enzyme loci along a transect through a single demographically characterized population: some loci exhibit clines, others complete substitutions, others uniform high polymorphism, and still others no variability (Johnson, 1975). The diverse microgeographic patterns seen at different loci within the same genetic population of butterflies suggest that adaptive complexes of particular alleles may be organized into metabolically coherent functional units, the optimal assemblage differing at different points along the transect.

It seems clear that a more pointed analysis is required, in which specific loci are chosen for examination because their enzyme products regulate the rate of key pathways of intermediary metabolism. Metabolic loci must be studied coordinately if we are to gain a clear picture of interactive effects.

THE HYPOTHESIS OF HOMEOSTATIC ADVANTAGE

From a physiological viewpoint, selection will act upon each step of a biochemical process as a function of how each affects the output of that process, the metabolic "phenotype." Those reactions which critically affect flux through pathways should be the most sensitive targets of selection. If enzyme polymorphisms reflect selection, then they ought to occur predominantly at rate-limiting steps (Johnson, 1974*b*). Fortunately a great deal of hard *in vivo* physiological data is available to test this hypothesis. Extensive comparisons in *Drosophila,* small vertebrates, and man (Johnson, 1974*b*) strongly confirm this prediction (Table II). Rate-limiting regulatory reactions consistently exhibit far more polymorphism than those which are nonregulatory.

These results argue against a random basis for the bulk of genetic variability; they also suggest a multilocus hypothesis which would produce generalized heterosis and thus account for the high levels of variability observed.

Considered from a functional viewpoint, metabolism in a variable environment presents a particularly difficult evolutionary problem: how to coordinate and integrate separate metabolic processes, each of which might respond quite differently to an environmental change. Thus, the rate-limiting enzymes of glycolysis and of pentose metabolism may respond quite differently to a change in temperature, while the need for coordination of these two processes remains. In considering metabolism from an evolutionary viewpoint, the problem of maintaining an integrated metabolism in a changeable environment is of central importance.

One cannot control a complex process without controlling the rate-

TABLE II. Metabolic patterns of polymorphism. Data are expressed as average heterozygosities. (From Johnson, 1974*b*)

Class of reaction	*Drosophila*	Small vertebrate	Man
Variable substrates	0.24	0.22	0.18
Specific substrates			
Regulatory	0.19	0.14	0.13
Nonregulatory	0.06	0.06	0.005
All loci	0.16	0.12	0.07

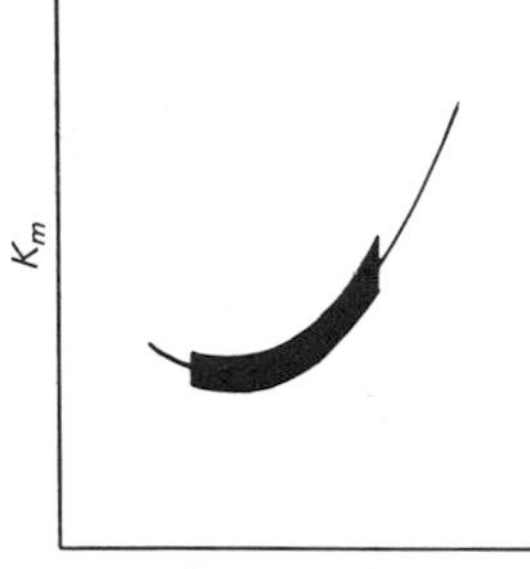

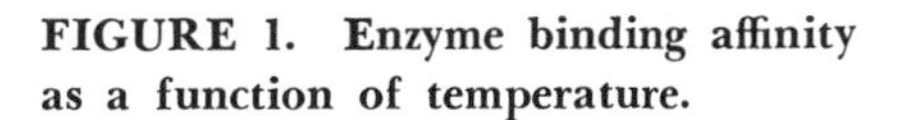
FIGURE 1. Enzyme binding affinity as a function of temperature.

determining elements. Whether building a house or running a metabolic pathway, the rate-controlling steps in the process must be buffered against random changes, or the process output cannot be regulated.

Enzymes at rate-limiting steps in metabolism are almost invariably under allosteric control. From what we have said about the fundamentals of protein structure, it is clear that a delicate balance is required to maintain both protein function and conformational sensitivity to allosteric effector molecules. Only over a rather narrow range of conditions (temperature, metabolite and effector concentrations, etc.) can a given protein maintain this balance. The response of such an enzyme to a change in temperature provides an example. In Figure 1 the binding affinity of a hypothetical enzyme is plotted as a function of temperature. The general form of the curve is typical of many proteins (Hochachka and Somero, 1973). The thick portion of the curve is drawn to represent the range of temperatures over which allosteric control might be expected. Below a certain temperature the molecule becomes too rigid, lacking the flexibility required for conformational change. The small amount of kinetic energy contributed by the binding of the effector molecule cannot overcome the energy barrier necessary to change the protein's conformation. At high temperatures the protein experiences the opposite problem. Thermal kinetic energies are so high that the protein's structure is very flexible and too disorganized for regulation by a small additional input of kinetic energy from the binding of the small effector molecule. Thus, allosteric sensitivity is possible only over a restricted range of conditions.

Regulatory loci thus present a critical problem in the physiological coordination of metabolism. The very properties which make them points of metabolic control also render them particularly sensitive to disorganizing influences from the environment.

However, a regulatory locus in a *heterozygous* individual presents

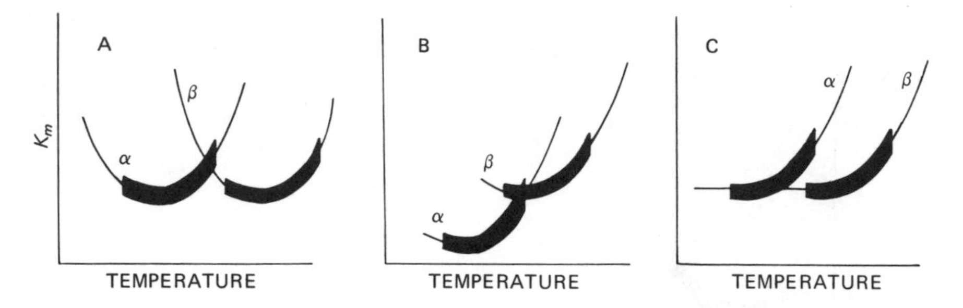

FIGURE 2. Mixed strategies of metabolic regulation. A: Negative thermal compensation. At low temperatures form α has the lower K_m and binds most of available substrate; both forms have allosteric sensitivity over the same range as their optimal affinity. B: Displaced affinity. Rather than form β having a higher K_m at low temperature, form α has a lower one. C: Differential turnover number. At low temperature α has a higher intrinsic rate and thus makes the predominant contribution to reaction rate, even though form β also binds substrate.

quite a different situation. Unlike its homozygous counterpart, the heterozygous individual contains *two* allelic forms of the rate-limiting enzymes. Available data indicate that these two allozymes usually differ in their kinetic responses to changes in reaction conditions. Figure 2 describes three such allozyme pairs. In each case *the multiple forms serve to expand the range over which coordination is possible,* primarily by minimizing variation in substrate affinity (K_m). The importance of a restricted range of K_m to metabolic control has been lucidly discussed (Hochachka and Somero, 1973). The different multiple forms act to buffer the regulation of metabolism against environmental variation.

The finding of high levels of heterozygosity at rate-limiting reactions (and very low levels at non-rate-limiting ones) now takes on added significance. In a typical physiological situation where substrate concentrations for metabolite-processing enzymes are low, a homozygous individual can maintain a regulated flux at a rate-limiting reaction only over a narrow range of temperature; in a heterozygous individual, however, reaction flux may be modulated over a far broader range of environments. This both suggests a generalized basis for heterosis at the highly polymorphic rate-limiting reactions and explains why non-rate-limiting reactions are generally not polymorphic.

The above hypothesis makes no assumptions about the nature of heterologous molecules. Hybrid multimers need not be superior in any way to maintain the heterosis and, indeed, may be inferior. The hypothesis outlined above suggests that we must look to the properties of the homologous proteins themselves and not to the heterologous multimers if we wish to understand the physiological implications of polymorphism.

This hypothesis, that enzyme polymorphism results from the homeostatic advantage of a mixed regulatory strategy, shares aspects of both the environmental-balance and the heterologous-advantage hypotheses. It involves both a variable environment and a molecular basis for heterosis. It is a multilocus hypothesis that will require a coordinated examination of metabolism for its evaluation.

The logical first step is for one to investigate the kinetics of alternative allozymes in a well-characterized natural population, such as the Alpine *Colias,* looking at enzymes catalyzing reactions critical to metabolic coordination and characterizing the binding kinetics of substrates, cofactors, and effectors as functions of environmental variables, such as temperature. One could then address directly the key question, which is whether a particular constellation of alleles can be shown to operate as a metabolic functional unit. If such is the case, adaptive alternatives must involve allelic substitutions at many loci. Selection would act to maintain entire integrated phenotypes. While physiological reasoning and recent field results make such a conceptualization attractive, there is as yet little or no biochemical data with which to evaluate it.

Ultimately this question must be addressed empirically *in vivo.* It will be necessary to assess directly the physiological consequences of an allelic substitution at an enzyme locus by monitoring the flux through that allozyme's reaction and through other related reactions under various environmental conditions. *In vivo* analysis under different conditions permits one to ask directly whether functional differences exist, whether such differences are physiologically significant, and whether heterozygosity at this locus in this organism constitutes a physiologically appropriate response to the environmental change. Appropriate methodology now exists for such an *in vivo* analysis, although little work has been carried out to date along these lines.

METABOLISM VS. ONTOGENY

It is important to realize that almost all loci scored routinely in electrophoretic studies of enzyme polymorphism code for enzymes involved in intermediary metabolism, i.e., in the processing and utilizing of energy. These are the day-to-day enzymes of the cell; their polymorphism may reflect an evolutionary strategy for achieving metabolic coordination in a varying environment. For most of these enzymes, physiological concentrations of substrate are usually low (on the order of K_m or less), and the reaction rates are limited by substrate or cofactor concentrations rather than by the amount of enzyme present.

For developmental enzymes and those involved in determining shape and form, the biochemical constraints are very different. These reactions often proceed at or near the enzyme's maximal velocity, in the presence of near saturating levels of substrate. Regulation of these reactions usually involves changes in enzyme concentration *per se*. Thus, differential synthesis, rather than rate modulation, is their typical pattern of regulation.

While there is little or no information on levels of genetic variability at developmentally important loci, one might expect from considerations such as discussed above that polymorphism would occur among the proteins regulating transcription rather than among the developmental enzymes themselves. We very badly need information on this class of enzyme loci; for as Wilson has pointed out, these are the reactions which affect the morphologies which we usually associate with the evolutionary process (Wilson, 1975; and Chapter 13).

CONCLUSION

Over the last 8 years it has become clear that electrophoretically detected enzyme polymorphism is widespread among natural populations. The notion that the bulk of this polymorphism is of adaptive value is accepted as a working hypothesis by many workers; although the point has not been definitely demonstrated, it is supported by many indirect arguments. The most broadly held view is that the polymorphism is maintained in natural populations by some form of balancing selection. I have outlined the two major mechanisms proposed to account for such balancing selection and the available evidence concerning each. I propose that the selection is not acting upon individual polymorphic loci but rather upon integrated metabolic phenotypes and, indeed, that enzyme polymorphism is selected specifically because it buffers that functional integration from environmental perturbation.

Allosterically regulated enzymes, critical to the coordination of metabolism, can operate effectively only over a limited range of conditions. Individuals containing multiple forms are potentially capable of responding appropriately to the binding of effector molecules over a broader range. This may provide the basis of the adaptive advantage of much enzyme polymorphism. This view places its major emphasis not on the function of individual enzymes *per se,* but rather on the regulatory process used to coordinate them.

This hypothesis, although consistent with available data and attractive from a physiological viewpoint, remains untested. Its evaluation will require pointed multilocus analyses which fully document functional differences between allozymes. Study of allelic variation at the loci of physiologically coordinated reactions, specifically assessing differences in biochemical functioning, is of paramount importance. Not only will it permit direct evaluation of this hypothesis, but, far more importantly, it will generate a functionally oriented body of data badly needed in the testing of any future hypotheses concerning genetic polymorphism and enzyme function.

SUGGESTED READINGS

The biochemical correlates of adaptation are discussed in P. Hochachka and G. Somero, *Strategies of Biochemical Adaptation*, Saunders, Philadelphia, 1973.

Relationships between level of polymorphism and enzyme role are considered in the following articles: J. H. Gillespie and K. Kojima, The degree of polymorphism in enzymes involved in energy production compared to that in nonspecific enzymes in two *Drosophila ananassae* populations, *Proc. Nat. Acad. Sci. U.S.A.*, **61,** 582–585, 1968; F. J. Ayala and J. R. Powell, Enzyme variability in the *Drosophila willistoni* group. VI. Levels of polymorphism and the physiological function of enzymes, *Biochem. Genet.*, **7,** 331–345, 1972; and G. B. Johnson, Enzyme polymorphism and metabolism, *Science,* **184,** 28–37, 1974.

CHAPTER 4

ALLOZYME VARIATION: ITS DETERMINANTS IN SPACE AND TIME

MICHAEL SOULÉ

POPULATION SIZE: ITS RELATIONSHIP TO HETEROZYGOSITY

The neutralist paradigm (Kimura and Ohta, 1971*a;* King and Jukes, 1969), like the Hardy-Weinberg equilibrium of population genetics, can be used as a base line from which a perspective on empirical results can be obtained. Employing it in this fashion, one is able to predict that the proportion of loci in the average individual that should be heterozygous at equilibrium depends on N, the effective population size, and μ, the mutation rate to neutral alleles, in the relation $H = 4N\mu/(4N\mu + 1)$. If protein polymorphisms are by and large neutral, there should be a simple S-shaped relationship between H and N if one assumes equilibrium conditions only and that μ is independent of N (Figure 1; and Lewontin, 1974). The correction of this formula for electrophoretic variants (Ohta and Kimura, 1973) hardly changes the picture at $H < 0.25$ (Figure 2).

It is not an easy matter to examine the correlation between population size and heterozygosity. Although N is the most important variable in population genetics theory, it is the least amenable to quantification in the real world. The population size of a species is a complex concept. In fact a species does not have just one population size; it has several. There is the simple total number of individuals in the species. Then there are a variety of *effective* population sizes depending on *structure,*

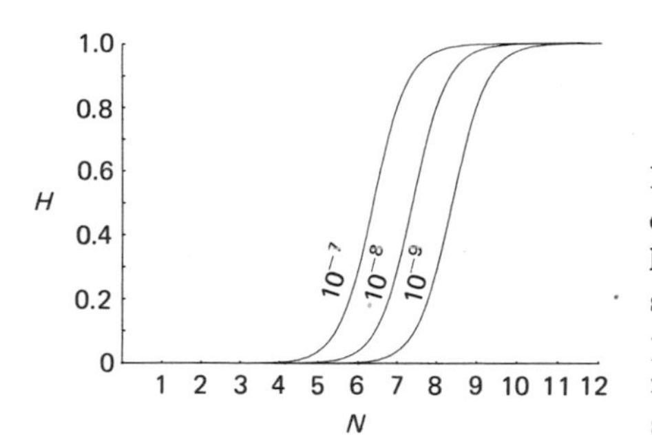

FIGURE 1. The neutral mutation-equilibrium relationship between heterozygosity (*H*) and population size (N, shown logarithmically as powers of 10) if one assumes different mutation rates to neutral alleles as shown.

i.e., the amount of inbreeding and the amount of isolation between subpopulations. Which among these different kinds of population size estimates should theoretically be correlated with the average heterozygosity?

The answer depends in part on what we mean by heterozygosity. Assume that a species has been sampled and electrophoretically surveyed at two widely separated localities. Further, assume that the observed heterozygosities in the two localities are 0.05 and 0.07. One way of calculating the heterozygosity for the species as a whole is to take the average of the two estimates, or 0.06.

Another way to calculate the heterozygosity for the species as a whole is to pool the data from the two populations and then calculate the heterozygosity as if there were only a single population. This kind of estimate could be much higher than the former, "observed," heterozygosity. For example, if for a particular locus different alleles are fixed in the different subpopulations, the "observed" heterozygosity for the locus is zero, but the pooled heterozygosity would be 0.50 for equal sample sizes. Pooled heterozygosity is an artificial concept, but it does convey something about potential variability in the species as a whole.

It is obvious that if we were using pooled heterozygosity estimates, we should use an estimate of the effective population size for the whole species. Another justification for using this population size estimate

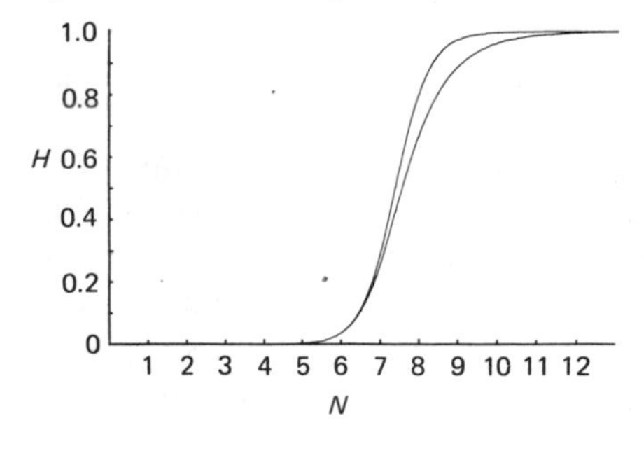

FIGURE 2. Same as Figure 1 except that the right-hand curve represents the relationship $H = 1 - 1/\sqrt{1 + 8N\mu}$; the mutation rate μ in both curves is equal to 10^{-8}.

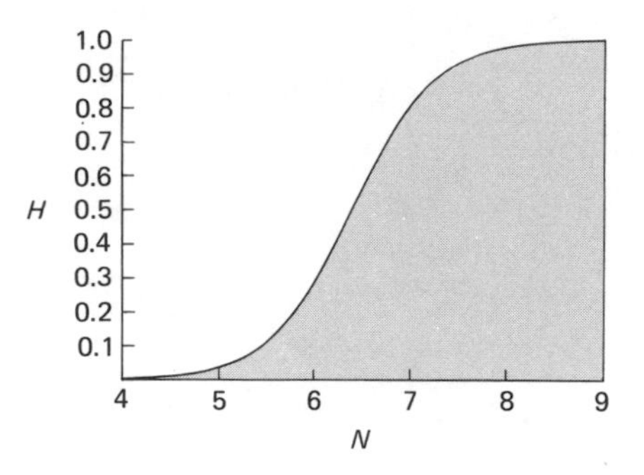

FIGURE 3. The region where all *H* values should fall according to the neutral theory. The curve is the equilibrium curve as in Figure 1 with $\mu = 10^{-7}$.

exists if the structure of the species approaches effective panmixis by virtue of the vagility of its members. One generalization that can be made from the electrophoretic data with considerable assurance is that species with excellent dispersal abilities, such as *Drosophila* or marine animals with pelagic larvae, have very similar gene frequencies throughout their ranges whereas animals with poor vagility manifest obvious geographic differentiation. I suggest that good dispersers approach effective panmixis; for them the appropriate N could be estimated by the actual number of adults in the species. For poor dispersers, the best estimate of N should lie somewhere between the subpopulation size and the species size. Exactly where depends on how isolated the subpopulations are.

Even if we are able to arrive at rather crude order-of-magnitude estimates of size or effective size, we are rarely able to assume that violent fluctuations of N have not occurred in the geologically recent past. For example, a bottleneck in N that occurred 10^4 generations ago could have a marked influence on H today. So, it would be very surprising indeed if a significant correlation were to be found between estimates of N and H. One might predict that the best one could hope for would be a relationship such as that shown in Figure 3. Some populations might lie on the curve, but many would fall below it because of the occurrence of population bottlenecks.

Mentioning the past has, perforce, allowed another variable to creep into our consideration—time, or better, "time since bottleneck," T. The less important time is, the greater should be the density of points on or close to the line in Figure 3. That is, if bottlenecks are relatively frequent and if the effect on H is long-lasting, then many or most populations will be far from their equilibrium H.

In Figure 4 I have attempted to estimate the effective population sizes of many animals for which heterozygosity data are available. These crude estimates are based on density, geographic range, and vagility. In all parts of Figure 4, I have assigned each species to a population size category. For lizards there are three categories. Category I ($N < 10^5$) includes burrowing species with no above-ground dispersal and with

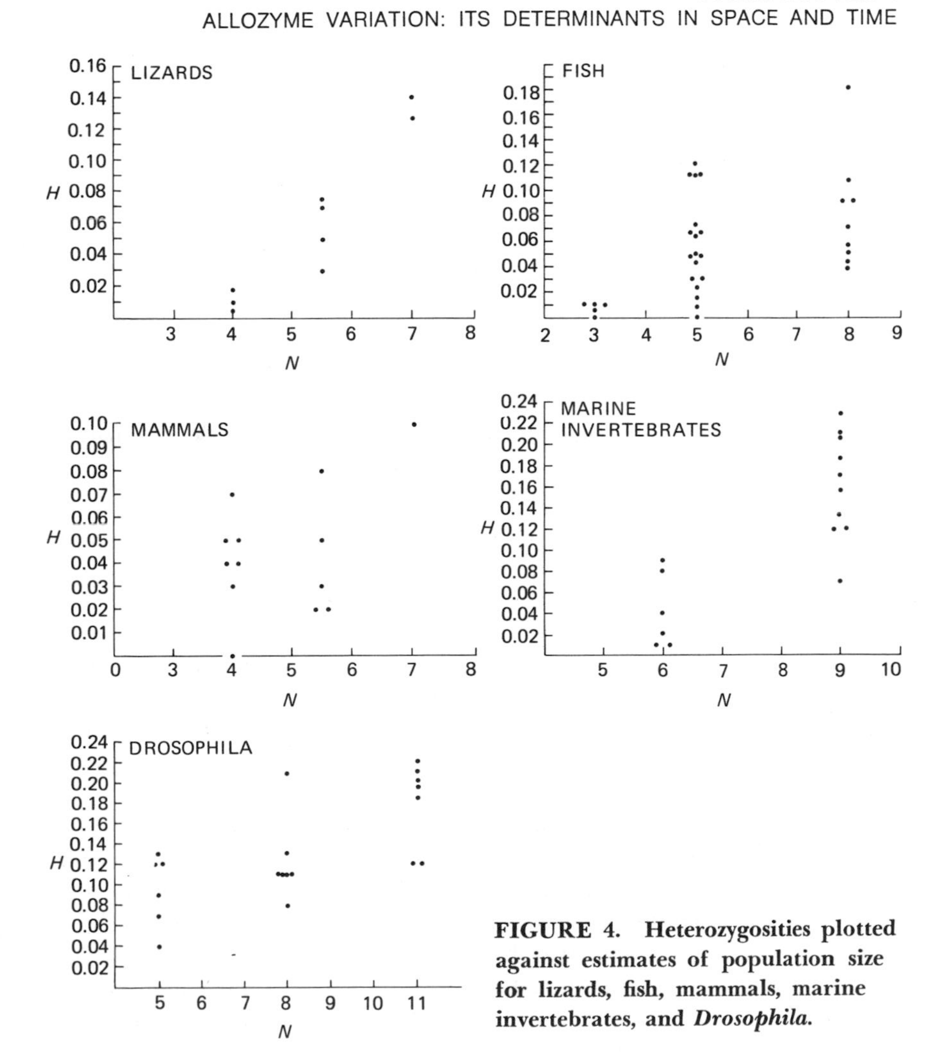

FIGURE 4. Heterozygosities plotted against estimates of population size for lizards, fish, mammals, marine invertebrates, and *Drosophila*.

much geographic variation. Category 2 ($N = 10^5$ to 10^6) includes relatively sedentary forms. Category 3 ($N = 10^6$ to 10^8) includes wide-ranging forms, or territorial species that are known to disperse long distances in the juvenile or adult stages and which show little tendency to form geographic races. Several sources of error are inherent in this procedure. One of these is differences in the loci sampled. Whereas in one species most loci may be dehydrogenases, in another most may be esterases, transaminases, phosphatases, and the like. It is reasonable that such heterogeneity should militate against the discovery of patterns.

Clearly the predicted correlation exists for lizards. The pattern is

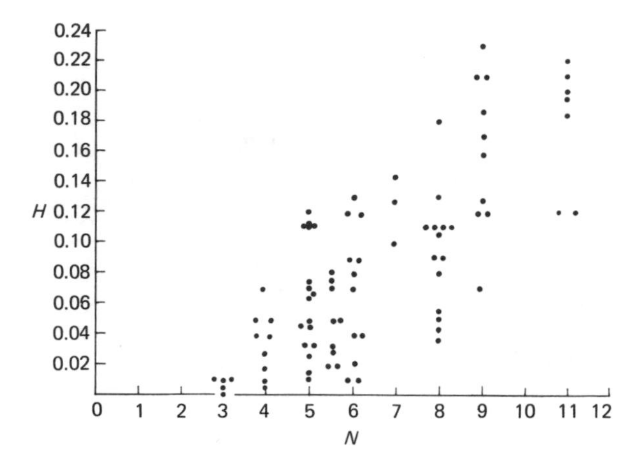

FIGURE 5. Summary of the data Figure 4 suggesting a correlation between heterozygosity and population size for animals in general.

similar in fish, mammals, marine invertebrates and *Drosophila*. Putting all the data together (Figure 5) we see that even with very crude quasi-relative estimates of N there is obviously an association between heterozygosity and population size. This is not the first time such a relationship has been discussed. Inger *et al.* (1974) found such a correlation in some frog species. Others (Soulé, 1971; Snyder and Gooch, 1973) have pointed out a correlation between heterozygosity and dispersal ability, or gene flow. In virtually all animals, some vagility is *sine qua non* for large effective N.

How much of the variation in H might be attributable to variation in N? If all populations were identical for T (time since last bottleneck), we could estimate this; but they are not. Consequently we must be satisfied with a very conservative estimate because heterogeneity of T should lower the observed correlation between H and N. Noting that the correlation overall between H and N is about 0.70, we estimate that our time-confounded N "explains" about half ($r^2 = 0.49$) of the variation in H. A much higher value might be found if T estimates were available for all species.

How far does the observed pattern depart from the neutralist model? First, in Figure 6 we superimpose our observed results on the theoretical curves relating H to N for a family of mutation rates to neutral alleles. Reasonable estimates for μ are between 10^{-7} and 10^{-9} (King and Jukes, 1969). If the upper boundary of the observed heterozygosities represents equilibrium populations, it seems that the neutralist prediction is not upheld—small populations have too much heterozygosity, and large populations have too little. No reasonable adjustment of mutation rates can accommodate the facts to the theory. Neutral mutation rates would have to span eight orders of magnitude! Spontaneous (Strickberger, 1968)

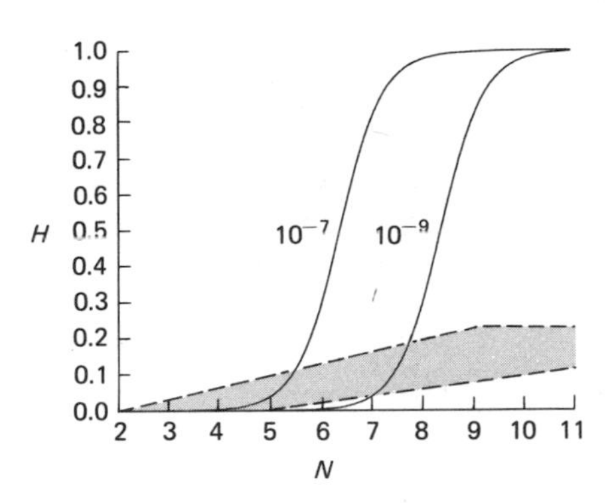

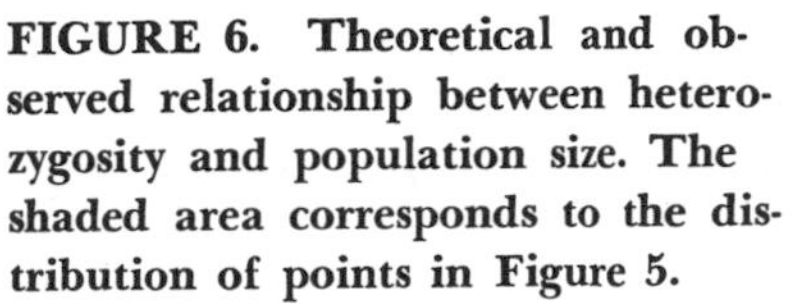
FIGURE 6. Theoretical and observed relationship between heterozygosity and population size. The shaded area corresponds to the distribution of points in Figure 5.

as well as induced mutation rates (Wolff *et al.,* 1974) do not vary over more than two orders of magnitude, at least in eukaryotes.

One assumption that would partially extricate the neutralist model is that only a fraction of the loci are potentially polymorphic. This would mean that the heterozygosity equilibrium would be at some value less than 1.0. This argument does not stand up to testing. It is indeed true that in a given species the proportion of loci that are polymorphic rarely exceeds one-half, but if we extend our vision to include a set of closely related species, it is clear that most if not all loci *can* be polymorphic. For example, in the *roquet* group of the genus *Anolis,* a monophyletic assemblage of nine species on the southern Lesser Antilles in the Caribbean, only two out of 19 loci studied were never polymorphic, and only three of these 17 polymorphic loci could be considered to be marginally polymorphic (from data in Yang *et al.,* 1974). This is not to say that all loci will eventually be found to harbor polymorphisms. None, for example, have been found in the glycolytic enzymes of the human brain (Omenn *et al.,* 1971). Nevertheless, the apparent upper limit of heterozygosity of 25 percent cannot be attributed to severe constraints on a majority of loci.

Another way to resurrect the neutralist model is to argue that it requires very long periods of time to reach equilibrium. For example, Ayala (1972) estimated that *H* for *Drosophila willistoni,* an abundant Neotropical fruit fly, should be over 0.99 assuming the validity of neutral mutation-equilibrium theory. He rejected the neutralist theory because the heterozygosity of *D. willistoni* is only about 0.18. Recently, Nei *et al.* (1975) have stated that this value of 0.18 supports the neutral theory assuming only that *D. willistoni* is a young species that passed through a bottleneck at its inception. This theory of the "recent bottleneck" undoubtedly can explain the low *H* values in some species. Without being diluvian, however, the bottleneck theory cannot be applied to all species. The most heterozygous species so far examined are cosmopolitan deep-sea invertebrates and the widespread tropical giant clam *Tridacna maxima* (Figure 4); their heterozygosities (0.23) are only slightly greater than that of *D. willistoni*. No known geological upheaval could be evoked to justify assuming bottlenecks in species such as these.

The results depart to a surprising degree from the neutralist base line. This is hardly cause for celebration, though, because there are a confusing array of selectionist theories all vying for the polymorphism pie. Before attempting to falsify some of these theories, I want to return to simple observations and ask some simple questions.

TIME: ITS RELATIONSHIP TO HETEROZYGOSITY

How fast does heterozygosity evolve? Michael Gilpin has kindly calculated the answer as it would be formulated under the neutral mutation model. Figure 7 presents his results based on the recent paper by Nei *et al.* (1975). Basically, Figure 7 shows that (1) small populations come to equilibrium faster than large populations but at lower levels of H and that (2) the rate of increase in H at any given time is proportional to population size. There are two problems, however: The first is that the evolution of polymorphisms depends on μ, the mutation rate to neutral alleles, and we do not know μ. What is worse, the neutral theory appears to be inapplicable to the real world, given the form of the relationship between N and H.

Fortunately, we now have some basis for an empirical attack on the rate of increase in heterozygosity. We have beeen investigating the evolutionary radiation of lizards of the *roquet* group of the genus *Anolis* on the Southern Lesser Antilles, Figure 8. The first colonization came

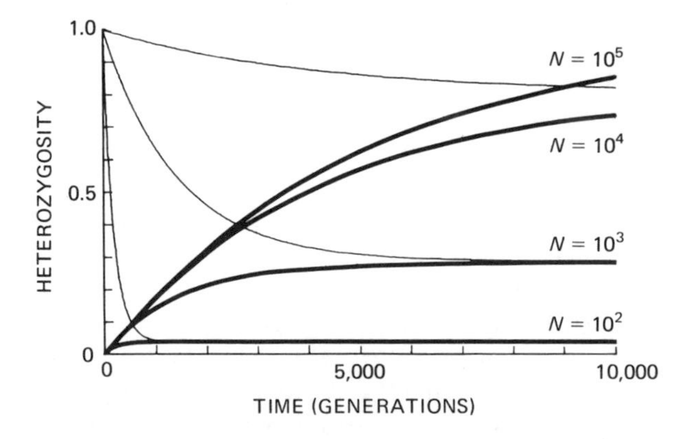

FIGURE 7. Heterozygosity for different effective population sizes as a function of time according to the neutral mutation-equilibrium theory. The ascending curves are heterozygosity; descending curves are homozygosity. $\mu = 10^{-3}$.

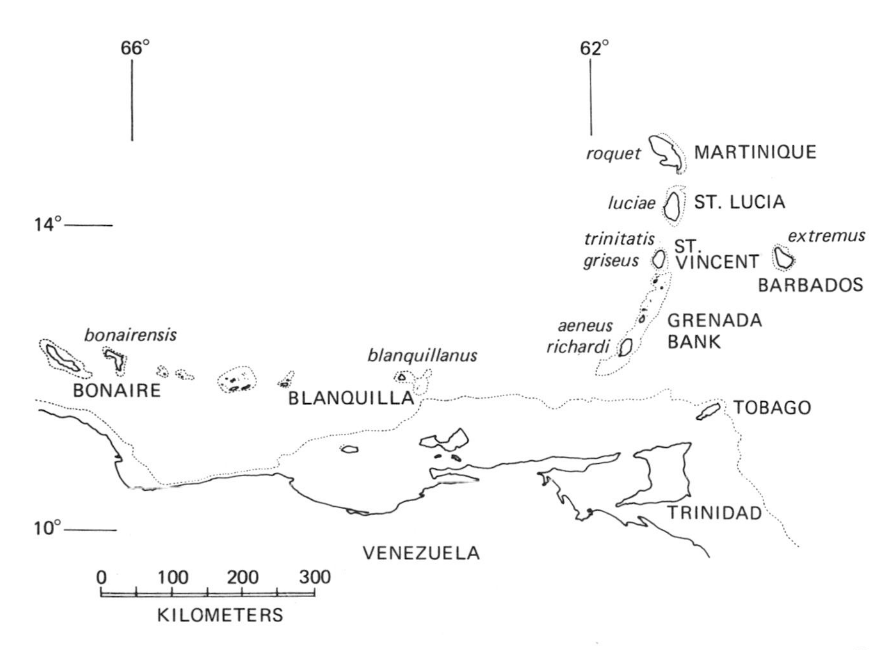

FIGURE 8. The distribution of the *roquet* group of the lizard genus *Anolis*.

from South America about 10 MY ago, and the colonization sequence and phylogeny of the group is quite well established (Yang *et al.*, 1974); see Figure 9. One of the nine populations, *A. luciae* on St. Lucia, was probably founded during the first stages of the colonization. It is the only anole on the island (until introductions in historical times) and retains many primitive characteristics including chromosome number, body size, and display behavior. I emphasize that this lizard has probably changed very little in the millions of years it has occupied this tropical island. It also has the highest level of H ($\approx$7.5 percent) found among the nine species. The population size is probably 10^6 to 10^7, and because St. Lucia is a high island, there is no reason to postulate any postfounding bottlenecks. I believe that it is safe to say that following a severe bottleneck at the time of colonization the polymorphism of *luciae* has been evolving for about 10^7 generations. Note that in generations this is 1,000 times longer than the evolutionary age of modern man, if one assumes he is about 100,000 years old.

If we assume that the heterozygosity of the source population was about 7 percent (typical of mainland iguanid lizards) and that of the founder population about 3 percent [the average effect of a bottleneck on heterozygosity is a 65 percent reduction (Nei *et al.,* 1975)], there has been an increase of about 5 percent in H per 10^6 generations or years in

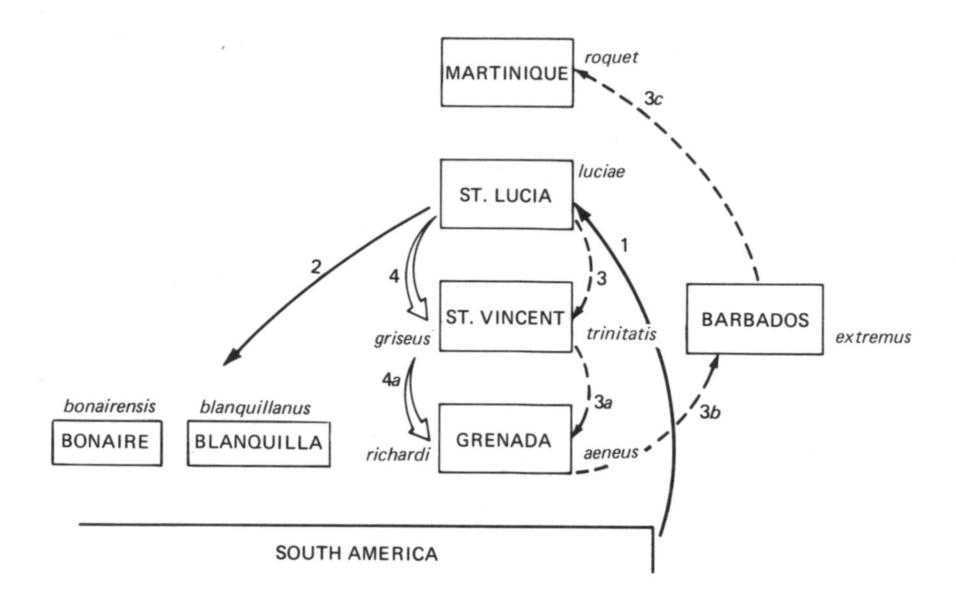

FIGURE 9. Hypothetical colonization sequence for the *roquet* group of the lizard genus *Anolis*.

this species. Such a slow rate of accrual of electrophoretic polymorphism is supported by data from other island reptiles.

In the entire *roquet* group, there are 76 alleles detected at 19 loci studied. Of these, 16, or about 21 percent, are unique, i.e., found in only one population at frequencies higher than 0.05. ("Unique" alleles at very low frequencies might simply have been missed in other populations because of sampling error.) In contrast to the plethora of unique alleles in the *roquet* group, there are virtually no unique alleles at frequencies greater than 0.05 in populations of lizards on continental islands ranging in age from 10,000 years to about 1 MY. This is true for *Uta stansburiana* on islands in the Gulf of California and for *Lacerta* species in the Adriatic (Gorman *et al.*, 1975). I emphasize that these relatively young island populations are not static evolutionarily—some of them are very distinct morphologically (Soulé, 1966). It would appear that electrophoretic polymorphisms in island lizards evolve very slowly, more slowly than morphology.

It could be argued that the rate of protein evolution is much higher at low heterozygosities and that it levels off as some optimum level is approached. An asymptotic decrease in rate may well be found; nevertheless it would probably be at a much higher H than 7 percent. In any case, the rate could not be very high at low levels of heterozygosity, or we would observe some unique alleles in the 30 or so populations of continental island lizards we have studied.

If such a slow rate of electrophoretic evolution is universal, then we might expect that the highest H would be found in environments conducive to extraordinary persistence and infrequent bottlenecking, such as the tropics and the deep-sea. Perhaps it is no accident, then, that among marine invertebrates the highest heterozygosities occur in deep-sea species ($H \approx 0.18$ to 0.22; Ayala *et al.*, 1975a,b) and the reef giant clam *Tridacna maxima* ($H \approx 0.23$; Ayala *et al.*, 1973). Incidentally, there are data suggesting that among fishes too the most polymorphic are those living in the deep-sea and in the tropics (Somero and Soulé, 1974).

Turning now to the distinction in H between invertebrates and vertebrates, one could argue that it could be explained just as easily by the time hypothesis as by ecological hypotheses. If paleontologists are correct in concluding that mammal species have an average persistence of only 1 or 2 MY (Kurten, 1968; Webb, 1969), significantly shorter than the average duration of about 6 MY for marine invertebrate species (Durham, 1971), then the often cited lower H's for mammals is explicable in terms of T, if one assumes only that bottlenecks or strong selection pressures are often associated with speciation events. Is the two-fold or threefold higher H value for invertebrates compared with rodents a consequence of the twofold to threefold greater age of invertebrate species? This might not be as far-fetched as it first appears. The generation time in rodents is often shorter than or equal to that of many aquatic and terrestrial invertebrates that reproduce only once a year.

Rates of evolutionary increase of heterozygosity on the order of 1 percent or less per million generations imply some interesting conclusions: First, because the average life expectancy of a species is between 1 and 6 MY (Stanley, 1975), most species cannot evolve an increase of more than a few percent during their tenure. Second, these low rates leave little room indeed for such subtleties as differences in niche width between closely related species to effect detectable changes in heterozygosity. Finally, time (since bottleneck) alone seems to be a very important source of heterogeneity in estimates of H. Millions of generations may be required to reestablish prebottleneck H levels. Even without considering time we were able to account for about 50 percent of the heterogeneity among species (the effect of N). With time, I hazard that we could explain much more.

EVOLUTIONARY RATE: ITS RELATIONSHIP TO HETEROZYGOSITY

One more factor, I believe, plays an important role in determining levels of heterozygosity. This factor is the amount of phenetic change,

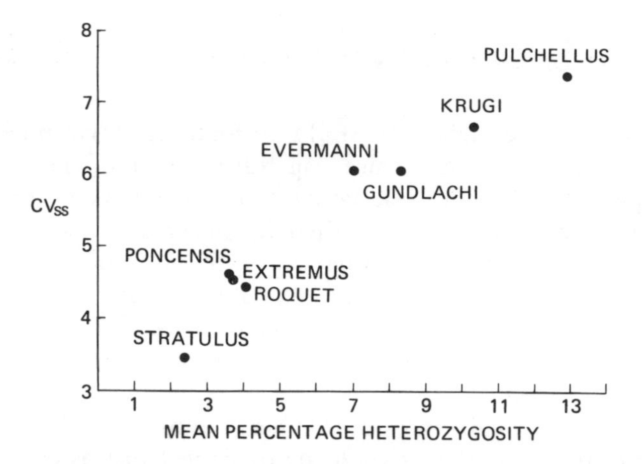

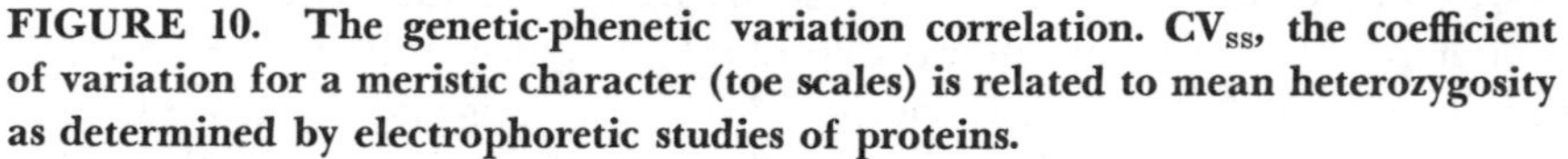

FIGURE 10. The genetic-phenetic variation correlation. CV_{SS}, the coefficient of variation for a meristic character (toe scales) is related to mean heterozygosity as determined by electrophoretic studies of proteins.

or average evolutionary rate. Elsewhere, I referred to it as the "divergence" part of the *time-divergence* hypothesis (Soulé, 1973; Soulé and Yang, 1973). As an example, consider *Anolis luciae.* This solitary anole has probably changed relatively little in the 10 MY or so that it has persisted. If the currency of divergence is genetic variation, then *luciae* has spent less of its genetic capital than any other species in the *roquet* group. In other words, it is the relative absence of divergence in *luciae* that explains why it has more heterozygosity than other members of the radiation that are just as old but have undergone much more phenetic change.

One of the assumptions implicit in this proposal is a correlation between H and the stuff of variation—additive genetic variation. That

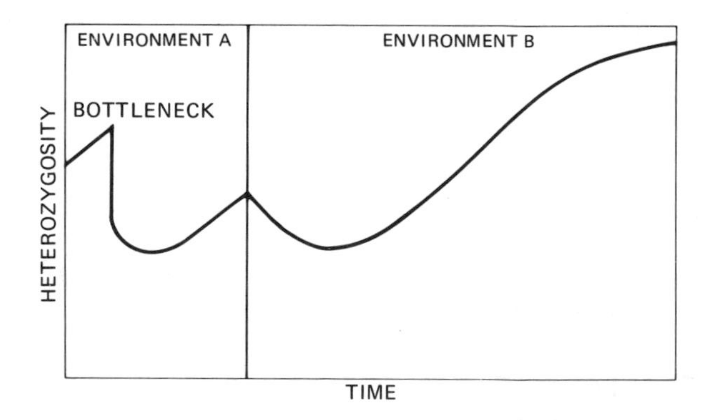

FIGURE 11. Hypothetical effect of directional selection on heterozygosity. Variation is lost upon exposure to the novel environment B; the rate of loss gradually declines as the population adapts to environment B.

is, this proposal assumes that the kind of genetic variation that is important in adaptation and is eroded by directional selection can be measured by variation in electrophoretically detectable proteins. Some evidence for such a correlation actually exists. At least there is an association between H and morphological variability for a scale character (Figure 10; Soulé *et al.,* 1973) among different species of island anoles. This correlation by no means necessarily implies that the stuff of evolution is polymorphism for enzymes. More probably, polymorphism at regulatory gene loci is the basis of variation in quantitative traits. The two kinds of genes (structural and regulatory), however, may both accumulate variation as a function of population size and time; so the two kinds of variation should also be correlated.

The erosive effect of directional selection on H is shown diagrammatically in Figure 11. Populations in relatively stressful, coercive environments could have a zero or negative rate of change in H. We might predict that lineages with high rates of evolutionary change will have relatively low H values (for a given N and T).

HETEROZYGOSITY AND NICHE THEORIES

Most investigators believe that there is a causal relationship between H and some type of environmental variation to which the population is exposed. Such theories can be classified into three categories: (1) the environmental-grain variation theory, (2) the environmental-amplitude variation hypothesis, and (3) the resource-predictability variation hypothesis.

The first theory is the oldest and most popular (Ludwig, 1950; Levene, 1953; Levins, 1968; Selander and Kaufman, 1973*a*). The proponents propose two kinds of species: (1) Coarse-grained species can exist in two or more niches, but individuals spend most of or all their lifetimes in only one of these. To the individual of a coarse-grained species the universe is relatively homogeneous; to the entire species it is a mosaic. (2) Fine-grained species are those in which the individual can exploit all or nearly all the patches in the environmental mosaic. It is argued that fine-grained species should have jack-of-all-trades alleles fixed at most loci whereas selection in coarse-grained species will favor different alleles in the different patches. Because of this, coarse-grained species will be relatively heterozygous. The higher heterozygosities on the average for invertebrates compared with vertebrates is marshaled as support.

If we try to test this theory within a somewhat more homogeneous taxonomic assemblage—vertebrates—it appears to fail. According to the grain theory, species with limited vagility, such as burrowing mammals, would have to be considered coarse-grained because different individuals are restricted to different patches of habitat during their lives. The grain theory predicts that these forms should be the most heterozygous species in their respective classes. The data in Figure 4 demonstrate that this is not so far true, because the burrowing species are if anything less polymorphic than other mammals.

Proponents of the second category of ecological hypotheses, the amplitude hypothesis, propose that variation in food, temperature, etc., within an individual lifetime should select for those individuals with the maximum biochemical flexibility; heterozygosity is supposed to confer this. This theory is different from the above because it requires only resource or habitat changes in time in contrast to patch differences in space.

Many studies could be mentioned that have been interpreted as tests of this niche-width variation hypothesis. The general finding is that widespread, vagile, or common species have higher heterozygosities than local, sedentary, or rare species. Therefore, a caveat must here be entered: Species differences in heterozygosity cannot be used to test the amplitude hypothesis unless population size and structure are controlled because N and T considerations would lead to the identical predictions as the niche-width hypothesis. To be explicit, species that are local, sedentary, or rare also have small N's; a high probability of bottlenecking, inbreeding or drift; and a high probability of a recent origin and founder effects.

Acknowledging the failure of the first two ecological theories to explain the available data, Valentine and Ayala (Valentine, 1971; Ayala and Valentine, 1974) suggest that predictability or dependability of food resources is causally related to heterozygosity. (See Chapter 5.) The hypothesis can be expressed as follows: Where resource (food) abundances are highly predictable, the best genetic strategy is to produce a variety of offspring which are genetically able to exploit different foods, thus minimizing intrasibling and intraspecific competition. In unpredictable circumstances, genetic uniformity is predicted because a generalist genotype is supposed to be better suited for coping with uncertainty. It is explicit in this theory that each of the highly heterozygous deep-sea invertebrates and tropical drosophilas, as well as *Tridacna,* is actually composed of many trophic specialists capable of finding the foods for which they are genetically predisposed. In other words, such species are actually genetically determined arrays of morphs, each morph exploiting a narrow range of the resource spectrum.

This theory is plagued by the same problems as the former two theories. The proposed genotype-environment interactions have never been demonstrated, nor is there any evidence for the trophic specialization that is the basic assumption of the theory.

Although this observer finds the correlation between niche variables and heterozygosity to be poor, others could counter that Figure 5 presents such evidence. It could be reasonably argued that the correlation between N and H is fortuitous, that it is the result of a correlation between population size and, say, niche width. Indeed, it is very likely that the geographic range of a species is more or less dependent on its physiological and behavioral flexibility. Population density, too, might often be related to ecological amplitude. And because geographic range and population density determine population size, the observed relationship between N and H might not be causal. I would, however, point to the obvious. It is inescapable that protein polymorphisms originate over long reaches of time, and the probability and rate of origination must depend on μ, N, and T. N and T, therefore, *must* explain some of the heterogeneity of H estimates. At the very least then we have shown that the niche-width variation hypotheses cannot be supported by studies that do not control for population size and time since bottleneck. This point is worthy of special emphasis when intraspecific comparisons are made. Many workers have attempted to explain the differences in H between populations of a species in terms of differences in niche width. One of the problems with this argument is that it assumes that heterozygosity can track geologically rapid shifts in resource availability, patchiness, or predictability.

If our empirical estimate of the rate that H increases is not off by more than a factor of 5 or so, it is virtually inconceivable that some wide-niched populations of a given species could evolve new polymorphisms that would affect relatively high levels of heterozygosity in the relatively short time usually available. It would require several hundred thousand to several million years for a wide-niched population to evolve a significantly higher heterozygosity than a narrow-niched population of the same species. At least in the Temperate Zone, however, the persistence of a population in the same geographic region and under the same ecological conditions for a fraction of this time is highly improbable. During the last million years or so, episodic glaciations and correlated climatic alterations have caused the range of virtually every Temperate Zone plant and animal species to repeatedly shift latitudinally and in many cases to split up and coalesce at least once. The possibility of tracking is not ruled out altogether, however. Changes in the frequencies of old alleles change heterozygosity levels. For example, if the rare allele at a locus with only two alleles increases in frequency to approximately 0.5, then heterozygosity will increase too. It is therefore conceivable that heterozygosity could track short-term environmental changes if selection significantly alters gene frequencies.

HETEROZYGOSITY: ITS REASON FOR BEING

In one sense the theory I am proposing here is a very reactionary one. It harkens back to Sewall Wright, Fisher, and classic population genetics. The essence of it is that differences in genetic variation at the biochemical level among species can be predicted from fundamental variables. N, μ, and T, and directional selection. Ecological niche theories may have some predictive value in ecology, but their predictive power in population-variation genetics is yet to be demonstrated. If ecological variation is not at the root of genetic polymorphism, however, then what is the motive power behind heterozygosity? Why does heterozygosity increase in lineages? The answer might be found in another kind of environmental heterogeneity—life cycle and tissue heterogeneity.

The suggestion is simply that a sufficient basis for selective maintenance of a polymorphism could be life cycle stage and tissue heterogeneity *within* an individual. Even in relatively constant physical and biological environments, such as the deep sea, this form of environmental heterogeneity exists. In a eukaryote there are inevitable physiological differences between larva and adult, placenta and fetus, or liver and brain. This is illustrated in Figure 12. Three genotypes for a particular locus are shown. The numbers beneath the genotypes are relative biochemical efficiencies rather than fitnesses as used in population genetics. (For example, the genotype *aa* is only 40 percent as efficient in the adult, or gut, as the genotype *Aa* and *AA*.) Averaged over all stages or tissues the heterozygote is superior to either homozygote, yet there is no heterosis *per se* in any single biochemical habitat. This idea is analogous to the concept of *marginal overdominance* (Wallace, 1968); but whereas marginal overdominance is a kind of apparent heterosis at the population level, the "ontogenetic" overdominance I speak of occurs at the organism level. Gillespie and Langley (1974) have recently refined this class of models. (See also Chapter 3.)

The above model is deficient in at least one respect; it cannot easily be tested. To reproduce *in vitro* the *in vivo* physiological environments relative to enzyme function and to test for the predicted allozymic functional differences would be a tour de force.

SYNTHESIS

By way of a summary, I will attempt to classify and relate the factors that determine heterozygosity in natural populations. Before these are listed, it should first be emphasized that the unit in which polymorphisms evolve is not necessarily the species (Figure 13). If the origin of the species is unaccompanied by a bottleneck and drift, then many of its polymorphisms would be inherited from its ancestor. The occurrence of shared electrophoretic polymorphisms in many species, e.g.,

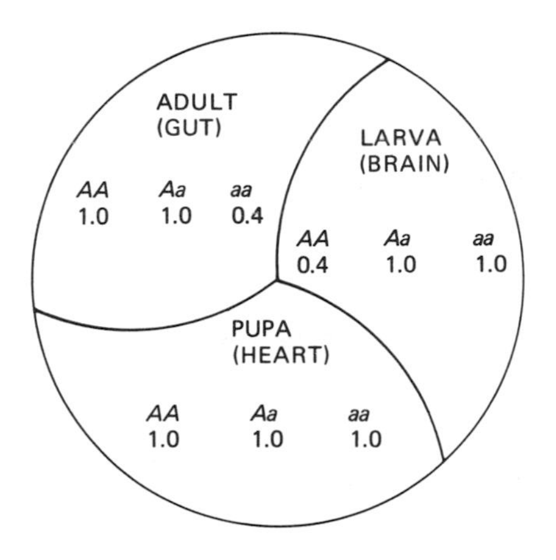

FIGURE 12. A model of the maintenance of a structural gene (protein) polymorphism due to life cycle or tissue heterogeneity. The numbers below the genotypes are hypothetical biochemical efficiencies.

*Drosophila willistoni (*Ayala and Tracey, 1974; Ayala *et al.,* 1974*a*), testifies to the great age of some of these and demonstrates that the lineage rather than the species is often the evolutionary unit.

According to the model proposed here, three conditions must be met before a population can have very high levels of heterozygosity. First, the population or lineage must be large; apparently only large populations can maintain high levels of polymorphism. Second, it must be old; new polymorphisms seem to appear at a very slow rate, so the number of generations since the last bottleneck is very important. Third, it must be evolving quite slowly because directional selection probably erodes heterozygosity.

What will be the nature and distribution of molecular polymorphisms in a sample of species taken at random? For simplicity, I have shown in Figure 14 species classified according to population, age, and

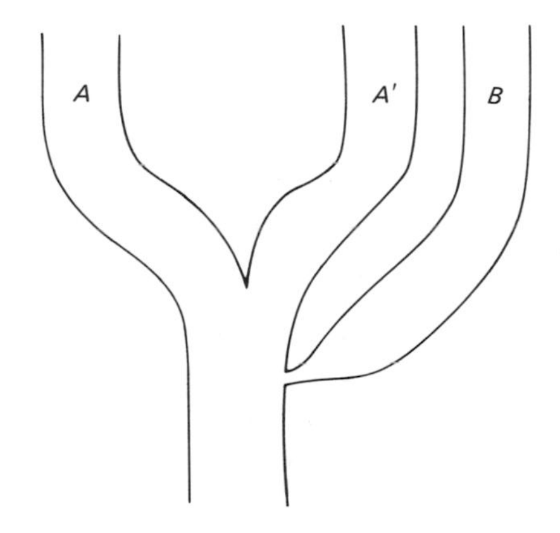

FIGURE 13. Many molecular polymorphisms are characteristic of a lineage. This is especially probable where a bottleneck in population size does not accompany speciation. Species *B* should share fewer polymorphisms with *A* or *A'* than either of the latter do with each other.

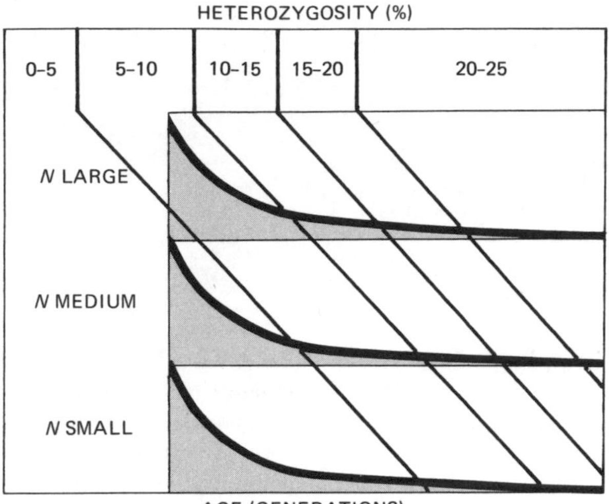

FIGURE 14. Hypothetical relationships between population size, the age of the population and heterozygosity. The negative slope of the curve relating frequency of species to age of species is assumed.

H. We reiterate that the essential feature of this model is that only very old, very large *N* lineages *can* have very high levels of polymorphism; they are the necessary conditions. However, not all large, old populations will be highly polymorphic, as is implied by the oversimplified figure. Those with high evolutionary rates may be losing variation as fast as or faster than they gain it.

The theory has at least one extremely disturbing feature—it can explain anything. Just as the invocation of natural selection can explain most biological phenomena, so can the invocation of the "time, size, and divergence model" explain any level of heterozygosity. Most of us prefer those theories that are easily falsifiable. This one is not. But even if it is unaesthetic, this theory may have some heuristic value.

SUGGESTED READINGS

Some of the source materials used for Figure 4 are the following:

Lizards: M. Soulé and S. Y. Yang, Genetic variation in side-blotched lizards on islands in the Gulf of California, *Evolution,* **27,** 593–600, 1973.

Fish: J. C. Avise and M. H. Smith, Biochemical genetics of sunfish. II. Genic similarity between hybridizing species, *Amer. Natur.,* **108,** 458–472, 1974.

Mammals: E. Nevo, Y. J. Kim, C. R. Shaw, and C. S. Thaeler, Genetic variation, selection, and speciation in *Thomomys talpoides* pocket gophers, *Evolution,* **28,** 1–23, 1974.

Marine invertebrates: F. J. Ayala, D. Hedgecock, G. S. Zumwalt, and J. W. Valentine, Genetic variation in *Tridacna maxima,* an ecological analog of some unsuccessful evolutionary lineages, *Evolution,* **27,** 177–191, 1973.

Drosophila: S. Lakovaara and A. Saura, Genic variation in natural populations of *Drosophila obscura, Genetics,* **69,** 377–384, 1971.

A summary of the arguments relating various environmental parameters to population structure is presented in chapter 4 of J. M. Emlem, *Ecology: An Evolutionary Approach,* Addison-Wesley, Reading, Massachusetts, 1973.

CHAPTER 5

GENETIC STRATEGIES OF ADAPTATION

JAMES W. VALENTINE

For mankind, the richness and variety of the Earth's environment contribute heavily to the pleasures of life. For organisms in general, this variability poses problems of adaptation. Each species possesses a limited range of environmental tolerances and requirements beyond which it cannot survive. Even within the range of tolerance, physiological and ecological responses and fitness may vary with local variations in the environment. Adaptations employed to cope with patterns of environmental variability have been characterized as *adaptive strategies* (Levins, 1968).

Strategies to cope with environmental changes through time permit species to persist through inclement conditions as well as possible and to exploit advantageous conditions as fully as possible. Probably the most important source of temporal variations to which species are adapted is seasonality; solar radiation, temperature, and a host of associated parameters are involved. In the sea, the tides form an additional important temporal variable, and of course diurnal variations occur in most environments. Many temporal variations consist of events that are irregular in amplitude and episodic in occurrence, such as weather conditions, and are therefore not very predictable.

Strategies to cope with environmental variability in space are chiefly concerned with maintaining population densities, dispersion patterns, and living ranges or areas that are adequate to support reproductively effective populations. If an environment is highly heterogeneous or patchy spatially, then species may have to live in many different patch types in order to maintain appropriate population sizes and densities.

The concept of environmental grain (Levins and MacArthur, 1966; Levins, 1968) permits one to express the relative environmental variabilities as they are perceived by organisms. If an environment is spatially fine-grained, then an organism may range anywhere as if the region were homogeneous; any patchiness is small relative to the tolerances of the organism. In a coarse-grained environment, however, an organism is greatly affected by the differences between patches, and in the most coarse-grained environment it may be restricted to a single patch throughout its entire life. The same environment may be coarse-grained for one species (the more specialized) and fine-grained for another (the more generalized). On the other hand, environments in which patches are most distinctive will tend to be perceived as coarser-grained by more species than will environments that are relatively homogeneous, other things being equal.

Organisms that perceive their environment as temporally fine-grained may treat it as if it were unchanging; for example, they may breed at any season. In a temporally coarse-grained environment, however, conditions are perceived as varying so much that adaptiveness is strained at times, and the most coarse-grained environment can only be inhabited periodically. (This usage differs somewhat from that of Levins, 1968.)

Patterns of variability differ from one environmental parameter to the next, on scales ranging from local to global. In the marine realm, for example, temporal variability in shallow-water temperature is greatest in midlatitudes and least in very low and very high latitudes, while variability in solar radiation is least in low latitudes and greatest at the poles. Furthermore, an organism may be tolerant of variation in one parameter but be restricted to a narrow range of variation in another, and thus different adaptive strategies may be developed for different parameters. Therefore the pattern of total environmental heterogeneity does not necessarily correspond to a pattern of adaptive strategies. Indeed, among the first questions we must ask is which parameters are most important in regulating adaptive strategies, and how is variability in those parameters to be judged.

Adaptive strategies have components, and these are often studied independently; strategies of feeding, of habitat, and of reproduction are common examples. It can be argued that many morphological characters are strategies (Vermeij, 1973), and it has been suggested that the critical morphological features that form the basic ground plans of many phyla and classes arose as components of an adaptive strategy in order to cope with novel patterns of environmental variability (Valentine, 1975). Features of the genetic system that are adaptive to patterns of environmental variability can be characterized as *genetic strategies.* The questions considered in this chapter are whether genetic variability, as indicated by enzyme polymorphisms, is selected as an adaptive strategy, and what environmental parameters may be involved.

Much work on enzyme polymorphisms has been stimulated by speculations on the role of genetic variability in the waves of extinction that are documented in the fossil record. Bretsky and Lorenz (1969) suggested that these extinctions occurred among species that were genetically depauperate because of prolonged evolution in stable environments. When their environmental regimes finally changed, their genetic resources were inadequate to evolve the appropriate adaptive responses. The general questions of how genetically variable the species may have been in different sorts of fossil associations and how this may have affected evolutionary invention, diversification, and extinction are of considerable interest.

SIGNIFICANT PARAMETERS IN ADAPTIVE STRATEGIES

To begin a search for environmental regulators of genetic variability, it is reasonable to identify the environmental factors that are generally most instrumental in evoking distinctive adaptive strategies. These factors may then be studied to determine their possible roles in influencing the genetic component of the strategy of adaptation. Contrasts in the properties of species from regions of high diversity with those from regions of low diversity are commonly believed to be due to differences in adaptive strategies. The explanations of why these properties arose and how they affect diversity form a convenient source of hypotheses of adaptive strategy regulation. (See Pianka, 1966; Sanders, 1968; Valentine, 1972.)

The oceans are in many ways simpler than terrestrial environments. Such variables as frost and rainfall are absent, and most biologically important factors, such as temperature, vary through less extreme states and change at lower rates than on land. Hypotheses that involve numerous environmental variables should be easier to test using marine data; we shall therefore examine strategy models as they apply to the marine realm.

Age of environment

It has been suggested that because diversification is a time-consuming process, younger environments contain fewer species than older ones (Fisher, 1960; Simpson, 1964; Sanders, 1968). The refrigerated climates of high latitudes date only from the late Cenozoic; Cretaceous and early Cenozoic polar climates appear to have been considerably milder. Therefore, communities in higher latitudes may be less diverse, at least in part, simply because natural selection has not had time to develop a large number of species adapted to present polar conditions. We may modify this suggestion and propose that adaptive strategies for life in novel environments require time to develop and therefore that diversity is lower in younger environmental regimes.

Achievement of a diversity equilibrium level, if it exists, does doubtless require time, but there is evidence that the present marine diversity patterns do not simply reflect age of environment. First, cooling in the deep sea must have closely approximated cooling in high latitudes, in age and intensity, yet deep-sea communities are highly diverse (Hessler and Sanders, 1967; Sanders, 1968). Second, there is fossil evidence of latitudinal diversity gradients during times of warmer poles (for example, Stehli *et al.,* 1969). Third, subtropical and warm temperate climates have a continuous history throughout the Phanerozoic, yet the diversity gradient is well developed across these ancient climatic zones (Valentine, 1972).

Physical environmental stability

A number of workers have suggested that variations in physical factors such as temperature, frost, salinity, or storminess regulate diversity of species. A variety of reasons have been proposed. Sanders (1968) has suggested that physiological stresses are greater in the more variable regimes and that fewer species are adapted to increasing conditions of stress. Diversity trends should therefore be inversely correlated with trends in physical variability. Furthermore, predictable variations are more easily accommodated by adaptations than unpredictable ones; regimes that vary unpredictably should have the lowest diversity of all.

Predictions derived from this model do not fit the observed pattern of diversity. In the sea, the greatest ranges of temperature variation and variation in a host of associated factors occur in mid-latitudes, especially off northwestern America and in the Sea of Japan (Sverdrup *et al.,* 1942). Marine climates are actually quite stable in high latitudes, as in the tropics. However, diversity is not lowest in the highly variable Temperate Zone waters.

There is a theoretical basis for minimizing the role of physical stability in diversity regulation, and this bears heavily upon hypotheses of adaptive strategies. Imagine a curve indicating diversity during a period when species numbers are permitted to rise to a certain level but no higher. The equilibrium level represents the species capacity of the environment. Species capacities evidently vary from place to place, because diversity trends appear to be general and persistent. The factors that regulate the capacity must be consumable factors that are used up by species, so that these factors do limit diversity; they are diversity-dependent factors (Valentine, 1972, 1973). Temperature and similar physical factors are not used up; they are strictly diversity-independent factors. Indeed, if natural selection can develop a species that can tolerate the physical regime of a given environment, no matter how variable the regime may be, then the presence of that species should not prevent others from becoming similarly adapted. If natural selection can develop one such species, it should be able to provide any number. Accordingly we need not review hypotheses of diversity-associated adaptive strategies that rely upon diversity-independent factors. As far as diversity-dependent factors are concerned, two come immediately to mind: habitat space and trophic resources; they have long been identified as important density-dependent factors.

Spatial heterogeneity of the environment

It is frequently suggested that more species can be accommodated in an environment that contains more potential habitats. Habitat space can certainly be used up; and where it exists in great variety, we find greater diversity, other things being equal. Rocky shores support a greater species diversity than do sandy beaches in otherwise similar environments. However, environments with similar spatial heterogeneities support much different diversities in different regions; rocky shores are more diverse in the tropics than in high latitudes. Thus, additional factors must be operating.

Trophic resource levels

Several hypotheses of diversity regulation have invoked the level of food supply. Some correlate high diversity with high productivity (Connell and Orias, 1964; Tappan and Loeblich, 1973*a*,*b*). This is the general correlation on land where tropical forests are more productive than those in higher latitudes, yet in the sea the pattern is anomalous. Pelagic productivity is high in Antarctic waters and in parts of the subarctic North Atlantic and North Pacific, yet diversity is low there; while in the low-productivity regions of major ocean gyres diversity is high. Deep-sea benthic communities are highly diverse despite a lack of primary productivity and a low supply of trophic resources; indeed they are especially diverse in the most oligotrophic regions (Hessler and Jumars, 1974). One might postulate that diversity and productivity are inversely related (Margalef, 1968; Valentine, 1971), yet coral reefs exhibit both high productivity and diversity. The effect of trophic resource levels on the evolution of diversity remains to be demonstrated.

Trophic resource stability

The factor that best correlates with interregional diversity patterns, as far as data permit comparison, is the stability of trophic resources. On land and in shallow marine waters the correlation is with stability of primary productivity, while in the deep sea it is with stability of detrital food supplies. This point has been argued from different perspectives by several workers (Klopfer, 1959; Margalef, 1968; Connell, 1970; Valentine, 1972, 1973). Latitudinal gradients in the stability of trophic resources are associated with the seasonality of solar radiation and therefore of primary productivity. In the sea, longitudinal patterns of productivity in shallow water are associated with seasonality or at least with episodic variation of nutrient supplies; regions with seasonal or intermittent upwelling or with seasonal alteration of nutrient-rich with nutrient-poor water types have lower diversities within comparable communities than latitudinally comparable regions of nutrient stability (Valentine, 1971). On land, the picture is complicated by frost, rainfall, and humidity regimes, but there does appear to be a relation between resource stability and diversity patterns.

An explanation for this correlation can be formed in terms of adaptive strategies and environmental grain. When the temporal grain for trophic resources is perceived as coarse, then the spatial grain tends to be perceived as fine. Where food is highly seasonal, a useful strategy would be to eat as many food items as possible, including living plants and prey as well as dead items or disseminated detritus and bacteria. This would certainly improve chances of finding food, even stretching the food supply well beyond the time of productivity. A correlated strategy would be to live in a wide variety of habitats. This would increase the species population density and range, heightening the chances of occurring in a region where food happened to be most persistent and, with a large population, increasing the probabilities of survival of enough individuals for successful reproduction following any heavy mortality because of an occasional prolonged inclement season. In other words, great temporal variability in food inhibits specialization, both for food items and for habitats. On the other hand if the temporal grain for trophic resources is fine, then species may perceive the spatial grain as coarse. The restriction to specialized habitats is visualized as the outcome of competition, with the partitioning of habitats and usually of food resources as well as a result (Dobzhansky, 1950). The stability of the food supply would permit the persistence of species despite restrictions of diet and habitat and the consequent lowering of population size.

In communities where productivity is highly seasonal or periodic, the argument goes, populations require a large share of the resources to sustain themselves, and relatively few species may share the resource base. Where productivity (or other resource supply) is stable, however, the resource can be highly partitioned, and diversity may be much higher. Presumably diversity will be highest of all in an environment with the greatest trophic stability and highest habitat heterogeneity. In the sea, coral reefs best combine these factors; they are legendary for the diversity of their biota.

There are a number of additional hypotheses of diversity regulation; especially notable are suggestions of predation (Paine, 1966) or disturbance, either physical (Dayton, 1971) or biological (Dayton and Hessler, 1972), as enhancing the diversity. The suggested principle is that these agents prevent resource monopolization by reducing the size of many populations, thus freeing resources for additional species. Because these agents are not diversity-dependent factors, we shall not consider them further individually. However, the possibility that diversity-independent factors could in fact control diversity deserves further comment. Even such a factor as temperature could conceivably operate to maintain a diversity level below the capacity set by diversity-dependent factors. Species are, after all, adapted to relatively narrow temperature ranges. If a sudden major change in temperature regime affected a large area, it could cause widespread extinction and lower the diversity accordingly. In time, immigration and speciation would act to develop an expanding biota of species adapted to the new regime. A second change, however, would again cause extinction. Continued changes in temperature (or other such factors) could hold diversity near some low level. Johnson (1970) has presented evidence that diversity-independent disturbances may control the spatial patterns of local diversities in Tomales Bay, California, while Grassle (1973) has interpreted some local diversity patterns on the Great Barrier Reef as resulting from storm disturbance. However, there is no evidence that the patterns of such sudden environmental disturbances do in fact correlate with regional patterns of species diversity. (Also, see above under "Age of Environment.")

CAUSES OF GENETIC VARIABILITY

With the advent of techniques of gel electrophoresis, large amounts of enzymatic variability, previously suspected only from less direct evidence, have been revealed in the gene pools of many species. The significance of the evidence of widespread polymorphism has been heavily debated. (See Chapters 2, 3 and 4.) It has been proposed (Kimura and Ohta, 1971*a*) that these polymorphisms are selectively neutral; they would thus all function equally well. However, evidence of the frequency and the patterns of the polymorphisms has failed to correspond with the predictions of the neutralist theory. These arguments need not be reviewed here; see, for example, Ayala *et al.*, (1974*b*). It now appears likely that the polymorphisms are due to some forms of balancing selection.

Levels of enzyme polymorphism are, of course, sensitive to some factors other than selection. In populations with reproductive systems that favor self-fertilization or have unusual mating ratios, genetic variability may be very low. North American populations of the self-fertilizing pulmonate gastropod *Rumina decollata* are monomorphic at all loci studied (Selander and Kaufman, 1973*b*). Populations that have developed from a few immigrants (such as is evidently true of American *Rumina*) or that have passed through size bottlenecks and then expanded from a restricted number of survivors may have low genetic variabilities because of a founder effect. Populations with small effective sizes may lose genetic variability because of gene drift. Isolated cave populations of the fish *Astyanax mexicanus* are nearly monomorphic (Avise and Selander, 1972), and genetic variation is low in small populations of lizards (two species of *Anolis;* Webster *et al.*, 1972) and of mice (*Peromyscus polionotus;* Smith *et al.*, 1973) on various small islands; these may reflect the effects of drift. In studying genetic variability as a possible adaptive strategy, cases where genetic variability reflects nonselective factors should obviously be avoided. Hence the following discussion is restricted to sexually reproducing, chiefly outbreeding species that are reasonably dense or widespread.

Among such species, the observed range of genetic variability is great in natural populations. Perhaps the best single measure of genetic variability is the percentage of loci at which an average individual is heterozygous (*H*, see Chapter 2). Throughout this chapter, genetic variability will be expressed in that manner. Estimates of average heterozygosity in marine invertebrates range from about 1 to over 20 percent (Table I). This large range requires explanation.

Age

Perhaps genetic polymorphisms tend to accumulate through time. In this event, older species should be genetically more polymorphic than younger. Soulé (1972; see also Chapter 4) in particular has proposed that in environments which minimize evolutionary change (interpreted to be tropical forests, coral reefs, and deep-sea environments) the older species should exhibit higher levels of protein polymorphism. Somero and Soulé (1974) present evidence that marine fish from the tropics and the deep sea tend to be genetically more variable than those from higher

TABLE I. Marine invertebrate species for which estimates of genetic variability are available that are based upon 15 or more loci

	Species	Individuals sampled per locus	Number of loci	Average percent heterozygosity
1.	*Asterias vulgaris* Verrill	19–27	26	1.1
2.	*Cancer magister* Dana	54	29	1.4
3.	*Asterias forbesi* (Desor)	19–72	27	2.1
4.	*Liothyrella notorcadensis* Jackson	78	34	3.9
5.	*Homarus americanus* Milne-Edwards	290	37	3.9
6.	*Crangon negricata* (Stimpson)	30	30	4.9
7.	*Limulus polyphemus* (Linne)	64	25	5.7
8.	*Upogebia pugettensis* (Dana)	40	34	6.5
9.	*Callianassa californiensis* Dana	35	38	8.2
10.	*Phoronopsis viridis* Hilton	120	39	9.4
11.	*Crassostrea virginica* (Gmelin)	200	32	12.0
12.	Asteroidea, four deep-sea species	31	24	16.4
13.	*Frieleia halli* Dall	45	18	16.9
14.	*Ophiomusium lymani* Thompson	257	15	17.0
15.	*Tridacna maxima* Roding	120	37	21.6
16.	*Euphausia superba* Dana	124	36	5.7
17.	*Euphausia mucronata* G. O. Sars	50	28	14.1
18.	*Euphausia distinguenda* Hansen	110	30	21.3

References: 1. Schopf and Murphy, 1973; 2. Hedgecock and Nelson, unpublished; 3. Schopf and Murphy, 1973; 4. Ayala *et al.*, 1975*a*; 5. Hedgecock and Nelson, unpublished; 6. Hedgecock and Nelson, unpublished; 7. Selander *et al.*, 1970; 8. Hedgecock and Nelson, unpublished; 9. Hedgecock and Nelson, unpublished; 10. Ayala *et al.*, 1974*b;* 11. W. W. Anderson, unpublished; 12. Ayala *et al.*, 1975*b;* 13. Valentine and Ayala, 1974; 14. Ayala and Valentine, 1974; 15. Ayala *et al.*, 1973; Campbell *et al.*, in press; 16. Ayala *et al.*, 1975*c;* 17. Valentine and Ayala, in press; 18. Valentine and Ayala, in press.

latitudes, although sample sizes are small and the ages of the species studied are not known. A difficulty with this hypothesis is that some very young species are known to display very high genetic variabilities. On the island of Hawaii, some endemic species of *Drosophila* are younger than 800,000 years, yet they are extremely polymorphic (about 17 percent; see Ayala, 1975).

Population size

Very small populations commonly have little genetic variability; therefore, it might be suggested that variability is correlated directly with population size. This would be the case under the hypothesis of selective neutrality among polymorphic proteins (Kimura and Ohta, 1971*a*). The data do not support this suggestion. Most of the largest populations studied thus far do have moderate to high levels of genetic variability. However, high levels are also found among relatively small populations. For example, the insular Hawaiian drosophila species must have vastly smaller numbers of individuals than members of many continental drosophila species, such as the widespread members of the South American *Drosophila willistoni* complex; yet on the average Hawaiian species exhibit high levels of genetic variability similar to species of the *D. willistoni* group (Ayala, 1975).

Geographic range

There is no obvious correlation between the extent of species ranges and their genetic variabilities. The similarity cited above between narrow-ranging insular and wide-ranging continental species of drosophila is a case in point. Nor are any correlations evident in the marine realm. On the contrary, a population of the essentially cosmopolitan deep-sea ophiuran *Ophiomusium lymani* has no more genetic variability than do other deep-sea forms that have far more restricted ranges (Ayala and Valentine, 1974).

Reproductive regime

It has been suggested that to search for possible correlations species with very different reproductive capacities should be studied in terms of their genetic variabilities (Ayala *et al.,* 1975*a*). There is as yet no strong evidence of such a relationship. Although precise data are still not available, it now appears that species which probably have greatly differing fecundities may have similar levels of genetic variability, even in the same environment. For example, the ophiuran *Ophiomusium lymani* is probably much more fecund than the brachiopod *Frieleia halli* (brachiopods tend to have low fecundities). These two forms are associated at many deep-sea localities. Both have rather high levels of genetic variability (*H* about 17 percent; Ayala and Valentine, 1974; Valentine and Ayala, 1974).

Physical environmental stability

Levins (1968) has concluded, largely on theoretical grounds, that genetic variability should be higher in unstable environments. From somewhat different reasoning similar conclusions have been reached by Bretsky and Lorenz (1969), Grassle (1972), and Johnson (1974*b*). The principle common to all approaches is that environmental heterogeneity, temporal or spatial, requires genetic variability to achieve tolerance to changes and to exploit the varied opportunities. This point of view is shared by many authors.

To test this hypothesis, one might study animals from stable environments to determine whether they have low genetic variabilities as predicted. In the sea, predictions of low genetic variability have been made on the basis of this hypothesis for animals from two specific environments, coral reefs and the deep sea (Grassle, 1973; Grassle and Sanders, 1973). Gooch and Schopf (1973) first studied deep-sea organisms from the standpoint of their genetic variabilities, and they showed that these organisms do not have particularly low levels as predicted. A series of later studies on several large deep-sea populations (Ayala and Valentine, 1974; Valentine and Ayala, 1974; Ayala *et al.*, 1975*a,b*) has revealed very high levels of variability, with heterozygosities about 17 percent and higher. Furthermore, populations of the tropical reef clam *Tridacna maxima* have been studied at Enewetak (Ayala *et al.*, 1973) and at Heron Island, Great Barrier Reef (Cambell *et al.*, in press) and have proved to be highly variable genetically (over 20 percent). Thus, specific predictions of this hypothesis have been falsified.

Mobility and size

Selander and Kaufman (1973*a*) have summarized electrophoretic data for a variety of animals and have pointed out that large, mobile animals (mostly vertebrates) have lower levels of genetic variability than small, relatively immobile animals (mostly invertebrates). Their interpretation is that this represents an adaptive strategy. The larger, more mobile organisms experience their environments as finer-grained than the small, less mobile forms. Therefore, for the larger forms the environment is less uncertain, and lower genetic variability is required for adaptation. (They include both physical and biological factors, such as

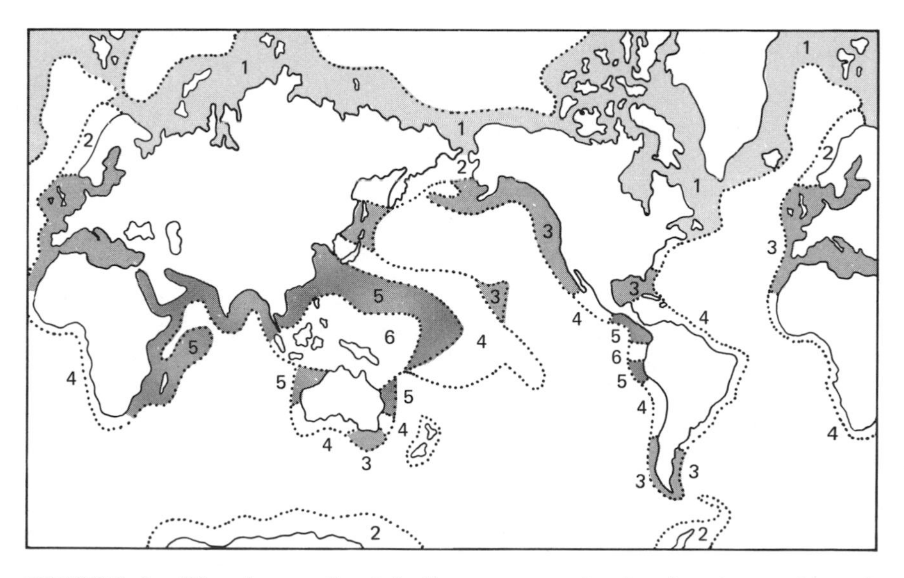

FIGURE 1. Diversity trends of shallow-water marine benthos (at continental-shelf depths) as indicated by Bivalvia, Gastropoda, Echinoidea, and Scleractinia. The lower the number in a region, the lower the estimated diversity of species in an average community. The deep-sea associations seem comparable to class 5. (After Valentine, 1973.)

food, in their concept of grain.) In the sea, there appears to be a similar trend, though data are few. Large, active forms (shallow-water asteroids and decapod crustaceans) tend to have lower levels of genetic polymorphism than do more sessile species from similar environmental regimes (phoronids, oysters).

Trophic resource stability

Trophic resource stability appears to be an important factor in those components of adaptive strategies that are associated with the diversity and structure of marine ecosystems (see above); hence, it is a plausible hypothesis that genetic strategies would also be correlated with food resource regimes. Quantitative data on resource regimes are not available for much of the ocean, nor has the available information been appropriately reviewed or synthesized. Therefore, relative species diversities have been employed as a first approximation to the relative stability of trophic resources. In Figure 1, the marine benthic animals of the world's continental shelf and shallow island regions, which include about 90 percent of all marine animal species, are classed according to their species

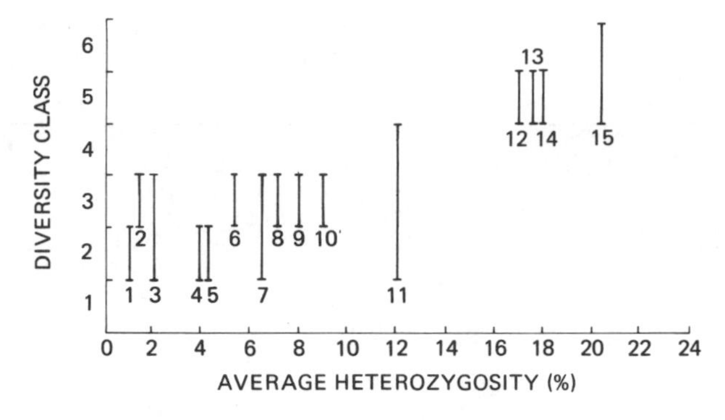

FIGURE 2. Genetic variability of benthic marine invertebrates plotted against trophic resource stability as estimated by faunal diversity. The diversity classes are those depicted in Figure 1. Species are listed by corresponding number in Table I.

richness. The resulting classes appear to be ordered according to their relative resource stabilities, where known. Class 1, in high Arctic latitudes, has the greatest seasonality of resources; class 2 (Antarctica and northwestern North Atlantic) has the next greatest; and so on. Class 6, chiefly the inner tropics of the west-central and eastern Pacific, has the most stable resource regime. In Figure 2, the average heterozygosities of benthic invertebrate populations are plotted against the diversity classes into which the species range today. The deep-sea species studied to date are not from the most diverse deep-sea associations. They have been placed in class 5, chiefly because Sanders (1968) has shown that deep-sea diversities approximate those of shallow-water benthic ecosystems of this class when similar (muddy-bottom) communities are compared. There is a clear trend of increasing heterozygosity with increasing benthic ecosystem diversity and presumably with increasing stability of trophic resources. The data of Somero and Soulé (1974) indicate that fish display a similar trend.

Only three invertebrate species have been studied from the pelagic realm (Valentine and Ayala, in press), but they form an interesting test of this hypothesis because they are closely related members of the same subgenus and have similar ecological functions but live in different resource regimes. They are species of krill (*Euphausia*). One lives in circumpolar water, one in temperate transitional water of the Peru Current, and one in Pacific equatorial water in the eastern Pacific. Because the diversity of pelagic ecosystems cannot easily be correlated with the diversity of benthic ecosystems, they are plotted separately (Figure 3). The Antarctic species faces the highest trophic resource seasonality and has the least genetic variability; the temperate form is intermediate in both respects; and the tropical species faces the least trophic resource seasonality and has the highest genetic variability. The observed genetic

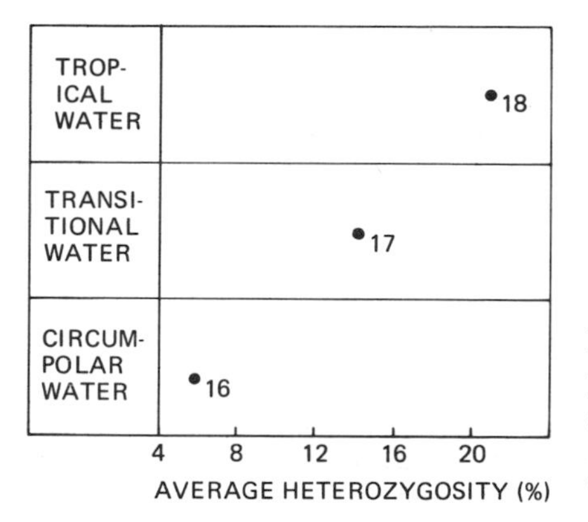

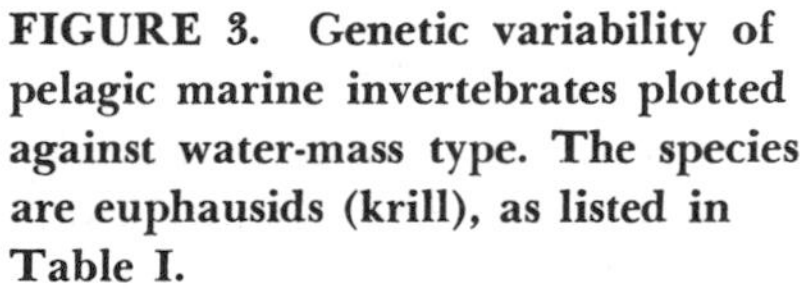

FIGURE 3. Genetic variability of pelagic marine invertebrates plotted against water-mass type. The species are euphausids (krill), as listed in Table I.

variabilities thus conform to predictions of the trophic resource hypothesis.

These diversity-correlated trends in genetic variability can be explained in terms of an adaptive-strategy model (Ayala *et al.*, 1975*a,b;* Valentine and Ayala, 1974). In environments which are temporally coarse-grained for trophic resources (such as high latitudes or regions of highly seasonal or intermittent upwelling), species pursue a relatively fine-grained spatial strategy. In order to perceive a heterogeneous environment as fine-grained, they must be flexibly adapted to tolerate a wide range of habitat conditions and to accept a wide variety of food items. Thus those alleles with products that function in the widest range of conditions are favored by selection; the gene pool therefore comes to consist of only the more flexible alleles, and only a few flexible genotypes are found. In environments that are temporally fine-grained for trophic resources (such as the tropics or the deep sea), species may pursue relatively coarse-grained spatial strategies and tend to be food or habitat specialists. This tendency must grow out of competition for limited (but stable) resources. However, alternative alleles which are slightly more advantageous in a certain restricted habitat condition or which confer an advantage to heterozygotes are permitted to accumulate in the gene pool. They are employed in adaptation to minor spatial variations in the environment, enhancing specialization. A great variety of genotypes are formed, many of which are highly fit in some special habitat condition.

A MOLECULAR MODEL OF GENETIC STRATEGIES

These data and their supporting arguments suggest a general model to account for the pattern of genetic variability observed among large

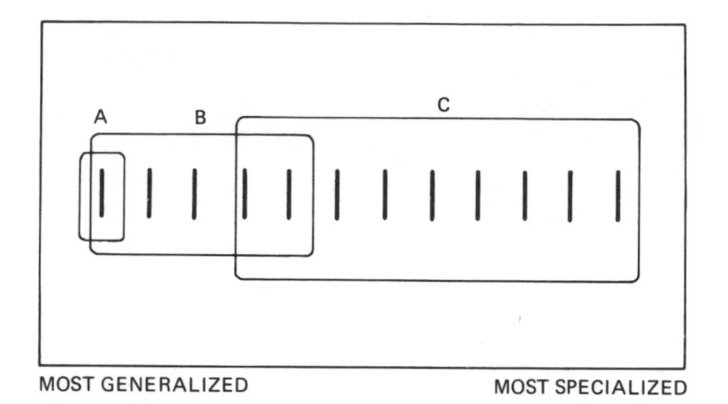

FIGURE 4. Hypothesis of selection for genetic variability in environments with different trophic resource stabilities. A: the allele with the most flexible functionally generalized product is favored by selection in the most unstable environment; B: a few alleles of intermediate flexibility are favored by selection in environments of intermediate stability; C: several specialist alleles are favored in highly stable environments wherein they promote adaptation to somewhat different microhabitat conditions.

natural populations of sexual outbreeding animal species. The model is most easily stated in terms of environmental grain. It is postulated that *species with fine-grained spatial strategies have less genetic variability than those with coarse-grained spatial strategies.*

The rationale is that fine-grained strategies require functional phenotypic flexibility. Figure 4 depicts for a single locus a number of functional alleles that appear at some rate in a natural population by mutation. They are positioned in order of the functional flexibilities of their products; proteins produced by those that lie to the left react with more substrates or are in some such way more flexible than those produced by alleles toward the right. The more specialized alleles toward the right may be hard to position properly because a given allozyme may function narrowly for one parameter but be slightly more flexible for a second while another allozyme may function narrowly for the second and slightly more flexibly for the first. If it is advantageous to follow a fine-grained strategy, then the more flexible allele will be selected—it may not function as well as a specialized allele in any one environmental state; but because selection is for overall fitness in the face of varied conditions, it will be favored. In species with coarse-grained strategies, specialized alleles which function better in a particular narrow environmental state will be selected. Patterns of variation of different parameters are commonly somewhat different, even on a small scale; thus, a variety of specialized alleles may be held in the gene pool by some form of balancing selection. Different allele combinations may be matched to appropriate combinations of environmental parameters.

A variety of predictions and speculations flow from this model. In general, any factors which affect the graininess of a species' adaptation should affect genetic variability. Trophic resource stability should regulate the modal genetic variability of species in a regional biota. Within a region, however, those species that are least mobile should have higher genetic variabilities. In the sea, these will commonly be suspension feeders and certain grazers. Detritus feeders and vagile carnivores should usually have lower variabilities, with large mobile predators having the least variation. Within any trophic category, the more specialized species would usually have the highest levels of genetic variability.

Genetic variability between separate populations of a species may have a different pattern from that between separate species. If genetic isolation is great enough, one might expect populations in the least heterogeneous environments to have the least genetic variability. The reasoning is simply that given a gene pool that contains a genetic variability and functional enzymatic breadth appropriate to the grain of the main portion of the species range, isolates will tend to preserve or eliminate alleles at polymorphic loci depending upon the heterogeneity of the local environment. If the environment were homogeneous with respect to some parameters, balancing selection might not occur for many alleles. This could explain the loss of genetic variabilities observed in cage experiments, which are greatest in the more homogeneous artificial environments (Powell, 1971; McDonald and Ayala, 1974).

Data from the marine environment tend to support this model, although many of the predictions have not been critically tested. As we have seen, trophic resource stability correlates directly with genetic variability. There is evidence that more active, mobile forms have less variability than more sessile species from similar environmental regimes, but more comparisons are needed. The terrestrial environment is much more complicated, and this makes it difficult to compare resource regimes. There does appear to be a suggestion of a latitudinal trend in genetic variability. Tropical species of *Drosophila,* for example, tend to have higher variabilities than do temperate or high-latitude species. The contrast between terrestrial organisms of different mobilities, such as between mammals and insects (Selander and Kaufman, 1973*a*), has already been noted.

When the model is applied to conditions in the geologically remote past, it gives rise to speculations (that are testable in principle) concerning evolutionary patterns. For example, waves of extinction tend to carry off the more specialized forms with coarser-grained adaptations, as far as we can tell. These should have been the most genetically variable species of their times. The remaining species had low genetic variability but would have possessed flexibly functioning alleles as well as flexible genotypes and opportunistic phenotypes. Perhaps such flexible forms provide better material for the creation of evolutionary novelties, which often appear following major extinctions.

With such a large variety of factors which could affect genetic variabilities as estimated by electrophoretic techniques, the broad correspondence of levels of polymorphism to the expectations of the adaptive-strategy model is promising.

SUGGESTED READINGS

A seminal work on adaptive strategies is R. Levins, *Evolution in Changing Environments,* Princeton University Press, Princeton, New Jersey, 1968. Levins argues that genetic variability should be higher in unstable environments. Similar conclusions are reached by P. W. Bretsky and D. M. Lorenz, Adaptive response to environmental stability: a unifying concept in paleoecology, *Proc. North Amer. Paleontol. Conv.* pt. E, 522–550, 1969; and by J. F. Grassle, Species diversity, genetic variability and environmental uncertainty, *Fifth European Mar. Biol. Symp.,* Piccin, Padua, 19–26, 1972.

Evidence and arguments relating genetic polymorphism to the stability of trophic resources can be found in F. J. Ayala, J. W. Valentine, T. E. DeLaca, and G. S. Zumwalt, Genetic variability of the Antarctic brachiopod *Liothyrella notorcadensis* and its bearing on mass extinction hypotheses, *J. Paleontol,* **49,** 1–9, 1975; and in J. W. Valentine and F. J. Ayala, Genetic variation in *Frieleia halli,* a deep-sea brachiopod, *Deep-Sea Res.* **22,** 37–44, 1974.

The ecology and evolution of marine invertebrates are comprehensively reviewed in J. W. Valentine, *Evolutionary Paleoecology of the Marine Biosphere,* Prentice-Hall, Englewood Cliffs, New Jersey, 1973.

CHAPTER 6

ORGANISMIC AND MOLECULAR ASPECTS OF SPECIES FORMATION

THEODOSIUS DOBZHANSKY

The species problem is the oldest problem in biology. Its roots go down deep in the prescientific era. Primitive man faced it when he invented names for animals and plants in his environment. It is symbolized in the Biblical story of Adam being called upon to give names to animals. In classical and medieval times the species concept was not altogether rigid: living beings were believed occasionally to arise from nonliving matter and be transformed into very different shapes. Linnaeus attempted to solve the species problem once and for all by supposing that each species was a product of a separate act of creation, but even Linnaeus in his late years somewhat relaxed the postulate of fixity and admitted that within narrow limits species can change and split. Darwin, as every student of elementary biology knows, concluded that species arise by gradual changes from ancestral species and may split into two or several new species. Of course, this did not close but rather opened the problem of species to scientific study.

NOMINALISTS AND REALISTS

Species, like subspecies, genus, family, etc., is a category of classification. Classification is a necessity. A biologist must be able to tell others what the organisms are he is writing or talking about. Is the species, then, only a device used to classify living beings? Ever since Darwin there have existed two schools of thought concerning the species problem. These are the nominalist and realist schools. Nominalists claim that species, like other categories, are man-made concepts, invented by classifiers for the pragmatic purpose of communication. The summit of the nominalist wisdom is this definition: "A species is what a qualified taxonomist considers to be a species." Individuals of a species have in common a name given by the classifier, and that is all.

Realists posit that species are phenomena of nature which exist regardless of whether they are recognized by human observers. A species is a form of integration of living matter; it is an array of individuals united by the bonds of sex, parentage, and descent. These bonds are, of course, less strong and obvious but nevertheless analogous to the ones that integrate the arrays of cells to become multicellular individuals. A realist wishes to make species the category of classification coincide with species the biologically integrated array of individuals and populations. This is desirable and convenient, but it is not absolutely necessary in principle. Species remain species despite some qualified taxonomists' refusal to recognize them. Thus, sibling species have all the biologically meaningful properties of morphologically distinguishable species, except that the former are identical or extremely similar in externally visible traits. Some conservative taxonomists have opposed formally naming sibling species. Their systematics remained incomplete, but sibling species refused to go away.

BIOLOGICAL SPECIES

Advances of evolutionary biology in recent decades have vindicated the realist position. The so-called biological-species concept was undoubtedly a step forward (Mayr, 1942, 1963; Stebbins, 1950). Among sexually reproducing and outbreeding organisms, species are, given adequate study, objectively ascertainable biological entities (Dobzhansky, 1935, 1937). They can be defined as Mendelian populations, or arrays of populations, which are reproductively isolated from other population arrays. Alternatively, a species can be said to be the most inclusive Mendelian population. Mankind and chimpanzee, *Drosophila pseudoobscura* and *D. persimilis,* lion and leopard, or the aspens *Populus tremuloides* and *P. grandidentata* are examples of closely related but discrete species entities.

The thesis of biological species brought forth its antithesis. In some particular situations the biological-species concept is difficult to apply. Certain authors hastened to conclude that the biological-species concept is invalid. The outcome of the confrontation was, however, not invalidation but development of the species concept in depth. It became recognized that not all species are stamped in the same mold. Speciation occurs in diverse ways, giving rise to diverse kinds of species, yet there is unity in the diversity (Mayr, 1970; Dobzhansky, 1970). Speciation is a sequel of sexuality. Whenever the reproductive biology allows any form of exchange of genetic materials between populations to take place, it is a biological necessity that the field within which the exchange occurs must be circumscribed. This is because exchange and recombination give rise to some well adapted and some ill-adapted genotypes. As the populations involved in the gene exchange accumulate genetic differences, the proportion of the well-adapted recombinants decreases, and the proportion of the ill-adapted ones increases. Natural selection restricts the gene exchange within tolerable limits. Speciation is an outcome of this restriction.

Sex is an ancient and venerable institution of the living world. So is species formation. The reproductive biology and the genetic population structure are, however, variable among diverse forms of life. The patterns of speciation conform to these variations. Obligatory cross fertilization is the most common method of reproduction in the animal kingdom, and it is not uncommon among plants. The biological-species concept has originally been framed for this situation, yet even among sexual outbreeders diverse states are met with.

VARIETY OF SPECIES

Consider the contrast between continental and island species. The former are often differentiated in geographic races connected by gene frequency gradients. The latter are separated by physical barriers which are only very rarely crossed. The five sibling species related to *Drosophila willistoni* are examples of continental species, and the "picture-wing" drosophilas of Hawaii studied by Carson are examples of island species. The former have differentiated presumably from allopatric races of an ancestral species and had to evolve a powerful reproductive isolation before they could coexist sympatrically (Spassky *et al.*, 1971; Dobzhansky, 1972). The speciation of *Drosophila* on Hawaii occurred chiefly by migration of single or very few founders from island to island. Postmating but no premating isolating mechanisms are characteristically observed among related species on different islands (Carson, 1971, 1973).

Animal species are, as a general rule, reproductively isolated from other species more rigorously than are plant species. It is among plants that most examples of interspecific hybridization, introgression, and hybrid swarms are found. Many animals produce few progeny per individual parent, and individual survival is hedged by physiological and developmental homeostasis. Some plants are long-lived and continue to bear large numbers of offspring for many seasons. Wide outcrossing and gene recombination "explore" the field of possible genotypes and may "discover" novel gene patterns of high adaptedness. However, "exploration" entails "cost." Many, most, or all new gene combinations may lack adaptedness. A more fecund species can afford the "cost" more easily than a less fecund one. Many plants and some animals are capable of asexual as well as sexual reproduction. The asexually, or parthenogenetically, produced progeny perpetuates genotypes of proved adaptedness; the sexual progeny may be devoted to genetic "experimentation" which generates novel genotypes (Stebbins, 1950; Dobzhansky, 1970).

A new species can be brought into being suddenly, as with fertile allopolyploids. More usual ways of species emergence involve slow accumulation of genetic changes, which climax in the development of reproductive isolation, making the divergence eventually irreversible. The gradual and slow transformation of an ancestral species into two or more derived ones has a consequence that is distasteful to a common type of mind which prefers having things either white or black and is repelled by grays. No definition of species can be framed that could be used to decide in all situations whether a given set of populations still belongs to a single species or to two or more distinct ones. Borderline cases are expected, and they are found. These are arrays of populations which are more distinct than races or subspecies usually are but less so than are most species.

Admittedly, borderline cases are nuisances to taxonomists, yet their existence was to Darwin, and continues to be, conclusive evidence that species have evolved, rather than having been brought into being suddenly. Gradual anagenetic changes of species in time create similar problems for paleontological taxonomists. Some paleoanthropologists attempted to escape the predicament by attaching species names to every scrap of fossil bone. The remedy was worse than the disease. *Homo sapiens* seemed to have an abundance of collateral relatives but no ancestors. Obviously one cannot prove that allochronic forms would or would not be reproductively isolated if they lived simultaneously. The only feasible solution is that suggested by Simpson (1953) among others —give species names to fossils that are as distinct from each other as average contemporaneous species of the same genus or family.

If speciation is a sequel to sexuality, one expects to find no species where the reproduction occurs without gene exchange between clones or pure lines. Nevertheless, the binary nomenclature is applied universally in the living world. In obligatorily asexual, parthenogenetic, or selfing forms, the species as a category of classification does not correspond to a biological reality of the same sort as in sexual outcrossers. Are then the nominalists triumphant over the realists? Only in part. In the first place, total lack of gene exchange turns out to be less frequent than it was believed to be (Jacob and Wollman, 1961). Transformation, transduction, and parasexuality go at least part of the way to fulfillment of the evolutionary function served by sexual outcrossing. In other words, new clones or pure lines are at least occasionally generated by gene change because of mutation, as well as by gene exchange because of sexuality.

DISCONTINUITY AND ADAPTIVE PEAKS

To say that there are no species except in sexual outcrossers is an overstatement, yet to expect species to be evolutionary units similar in kind regardless of reproductive biology is unwarrantable. What is more nearly universal than speciation is the discontinuity of organic variation. As pointed out by Wright in 1932, the existing genotypes of living beings are far from a random sample of the potentially possible ones. Instead, we find clusters and clusters of clusters of related genotypes occupying *adaptive peaks,* while the intermediates between the clusters are absent or extremely rare. Of course, the clusters of related genotypes give rise to clusters of recognizably similar phenotypes.

Systematists, from the primitive man through Linnaeus to Mayr and Simpson, put the clustering to use. Racial, specific, generic, familial, and other names were attached to clusters of various orders. This can be done about equally well with sexual and with asexual forms. There is however a basic difference. Although the clustering is a reality, choosing which clusters are to be classed as genera, families, or orders is arbitrary, or more precisely a matter of convenience. Until relatively recently two kingdoms, animals and plants, and a dozen or so phyla were distinguished in the living world; at present Margulis (1974) has five kingdoms and 89 phyla. This much nominalism wins. In sexual outcrossers there is, however, one kind of clusters which arises because of a basic evolutionary phenomenon—advent of reproductive isolation. This kind of clusters we call species. What is called a species in an asexual form is no more and no less arbitrary than what is called a subgenus or a genus or a family. But let it be emphasized that there is a whole gamut of reproductive biologies, starting with obligatory outcrossing (which is common) to obligatory asexuality (which is much less common). Parallel to this, there is a gamut of species situations, from strictly realistic to largely nominalistic.

HOW DIFFERENT ARE SPECIES?

Another basic problem which arises in any discussion of the species problem is how extensive are the genetic differences between species.

Most efficient methods to attack this problem are relatively new, namely examination of allozyme variants by means of electrophoresis, as discussed in Chapters 7 and 8 of this book. This is no reason to forget that the problem has had a long history and that meaningful findings had been made before methods of molecular biology became available. At the turn of the century de Vries (1901) believed that species arise by single mutational steps. This became untenable when Morgan (1919) and his collaborators observed many mutants arising in cultures of *Drosophila melanogaster*. Quite clearly, the mutants belonged to the same species as the parental form.

Yet Morgan, together with most evolutionists of his day, chose what became later known as the classical model of the genetic population structure as a working hypothesis. Individuals of a species were supposed to be alike and homozygous for a large majority of their gene loci; gene mutants, usually deleterious to their carriers and recessive, were responsible for the presence of rare heterozygous individuals. Most, though not quite all, species allegedly had a normal or *wild-type* condition, the deviations from which were discriminated against by natural selection. Acceptance of the classical model made almost inevitable the assumption that species differ in only few lucky mutational changes which happened to pass the scrutiny of natural selection. Not only closely related but even rather remote species are alike in most of their genes. Sturtevant (1948) and Muller (Muller and Kaplan, 1966) still adhered wholeheartedly to this classical model. Discovery of enzymes with identical functions in otherwise very unlike organisms, such as man and yeast, seemed to validate the classical model beyond reasonable doubt. It took more sophisticated studies to show that enzymes with identical functions are not necessarily identical in composition.

Evidence of a quite different nature came from studies of recombination in F_2 and later generations of fertile interspecific hybrids. Most of this work has been done with plants because fertile hybrids between animal species are rather infrequent. Baur (1925, 1932) described the hybrids between species of snapdragons, *Antirrhinum*. A spectacular diversity was observed among F_2 hybrids; there appeared not only all sorts of recombinations of traits of parental species but also traits present in neither parent, due evidently to interactions of the genes of the species crossed in novel combinations. Baur's minimal estimate of the number of gene differences between the snapdragon species was 100; this is evidently limited to genes which induce visible morphological differences. A segregating progeny of species hybrids growing together on the same field is one of the most impressive sights a geneticist can behold. No wonder some biologists were lured to rather extravagant speculations. Lotsy (1916) felt that evolution can be accounted for merely by recombination of some hundreds or thousands of preexistent but unchanging genes. Mutation was unnecessary.

SIBLING SPECIES

What about species, such as sibling species, which are ostensibly very similar or identical in appearance? Are they alike in all but a very small number of genes? Adherents of the classical model thought so. Similar mutants arise in closely related and in rather distinct species of *Drosophila*. If similar mutant alleles appear in different species, does it not follow that wild-type alleles of the gene loci producing these mutants are also similar? Experiments on sibling species disclose a more complex situation. *Drosophila pseudoobscura* and *D. persimilis* are virtually indistinguishable in appearance. Their F_1 hybrids are as vigorous as the parents, though the hybrid males are completely sterile. The backcross progenies suffer however a hybrid breakdown. Depending on the strains of the parental species used, a part, or even the entire progeny, fail to survive or appear as individuals with various abnormalities (Dobzhansky, 1936). Similar evidence was obtained by Pontecorvo (1943) and more recently by Weisbrot (1963) for *D. melanogaster* and *D. simulans*. Sibling species of *Drosophila* have clearly different genetic systems, which make them physiologically and ecologically distinct but leave the visible morphology unchanged. It is surely naive to expect that genetic divergence must be manifested uniformly in all kinds of characteristics. With the possible exception of the Hawaiian species, species divergence in *Drosophila* involves mainly physiological rather than morphological traits.

Erosion of the classical model began with the discovery by Chetverikov (1926) and others that natural populations of sexual outbreeding species carry innumerable genetic variants, mostly concealed in heterozygous condition. Far from being homozygous and alike, individuals in nature are all different and are complex heterozygotes. This is the foundation of the balance model of genetic population structure. The evidence in support of this model cannot be reviewed here. I limit myself to a single example, which in my opinion is impressive. This is the release of genetic variability by recombination of gene contents of chromosomes which are ostensibly alike in their effects on the phenotypes of their carriers (Dobzhansky *et al.*, 1959, and references therein). In four species of *Drosophila (pseudoobscura, persimilis, willistoni,* and *prosaltans)* chromosomes were tested for their effects on the viability of homozygotes. As expected, some were lethal, others semilethal, subvital, or quasinormal. Are the chromosomes of this last class the normal or wild-type postulated by the classical model, and are they all alike as the model leads us to expect? They are not. Females were made hyterozygous for pairs of these would-be normal chromosomes. Crossing over produced in the next generation chromosomes compounded of sections of the parental ones. Far from being all normal and alike, the chromosomes brought forth by recombination exhibited a range of viabilities from normal to lethal. Levene (Dobzhansky *et al.,* 1959) estimated from these results that recombination of the ostensibly similar and normal chromosomes re-creates in one generation a major part of the genetic variance present in the population.

Introduction of allozyme studies by electrophoresis opened a new era in evolutionary genetics (Lewontin and Hubby, 1966). This is not because this method uses cooking recipes borrowed from biochemists; rather the method overcomes at least in part an intrinsic limitation of Mendelian genetics. By observing segregations in hybrid progenies one can estimate the numbers of the genes in which the parents differ but not the numbers or proportions of similar genes. With the pioneer work of Harris (1966) and Lewontin and Hubby (1966), it became unambiguously clear that populations of at least the sexual outbreeding species are polymorphic at usually more than one-half of the gene loci and that an individual is heterozygous for 5 to 20 percent of its genes.

SELECTIONISTS AND PANNEUTRALISTS

The classical model is false. Its former partisans have made a clever about-face. The enormous amount of genetic variation now discovered in natural populations is biologically and adaptively insignificant. It is neither useful nor harmful; it is neutral. Lewontin (1974) has, in my opinion inappropriately, called this new panneutralist model neoclassical. The prefix "neo" does not do justice to the basic difference between the classical and the panneutralist models. The keystone of the former was the assumption of the prevalence of genetic uniformity and of normal or wildtype chromosomes and genotypes. Panneutralists do not deny the prevalence of polymorphism and heterozygosity; they merely assume it to represent a kind of noise in the genetic system (Kimura and Ohta 1971*b*).

The problem of the amount of genetic differentiation between populations, races, species, and genera can now be reexamined with the aid of methods vastly superior to the old ones, though still not free of serious defects. The old debates about species differing in many or in few genes were inconclusive because the best that could be done was to estimate the numbers of gene differences. Now estimates can be obtained of the proportions of loci represented by different as well as by similar alleles in populations of two or of several species. Such estimates have now been obtained for species in diverse groups of vertebrate and invertebrate animals, as well as of plants (reviews in Ayala, 1975; see also Chapters 7 and 8). The species of the *willistoni* group of *Drosophila* have been most thoroughly studied by Ayala and his students. The results of their work can be used as a paradigm for studies in this field.

The amount of genetic distance and genetic identity between populations can be measured with the aid of several statistical techniques. Ayala and his colleagues have chosen the method proposed by Nei (1972). Although Nei's original formulation involves the gratuitous assumptions that a majority of gene differences is adaptively neutral and that the accumulation of such differences is uniform and proportional to time, the numerical values of the genetic identity and genetic distance are not dependent on these assumptions. The value of genetic distance, *D*, can be interpreted more simply. Namely, it is a measure of the average number of electrophoretically detectable codon substitutions per gene locus which have accumulated since two populations have diverged from a common ancestor.

GENETIC DISTANCES BETWEEN SPECIES

The average genetic distance between local populations within a species and subspecies of the *willistoni* group of *Drosophila* is 0.031 ± 0.007; while between subspecies it is 0.230 ± 0.016, and between semispecies of *D. paulistorum* 0.226 ± 0.033 (Ayala *et al.*, 1974*c*). Local populations are geographically separate but not isolated reproductively. The subspecies used in this study are not only geographically separate but produce sterile male hybrids in one or both of the two possible reciprocal crosses. The semispecies are often sympatric and show a strong ethological isolation as well as a sterility of hybrid males in both reciprocal crosses. The estimates of the genetic distance between the semispecies is, thus, no greater than between the subspecies, despite an appreciable increase in reproductive isolation. This does not mean that the reproductive isolation has arisen without genetic divergence. What this means is rather that, given an appreciable genetic divergence, superimposition of reproductive isolation does not by itself demand very many additional gene differences. A genetic distance of 0.23, found between the subspecies as well as between the semispecies, suggests according to the interpretation indicated above that one or more codon substitutions have taken place in approximately 23 percent of the gene loci. Most people will agree that this is an impressive amount of genetic divergence.

Although sibling species of the *willistoni* group are scarcely distinguishable morphologically, the ethological isolation between them is very strong, yet not absolute, at least under laboratory conditions. Some authors claimed having obtained even fertile hybrids between the sibling species. The claims have not been verified. The genetic distance between the sibling species is 0.581 ± 0.039, more than twice that between semispecies. Finally, the genetic distance between morphologically distinct species of the *willistoni* group is 1.056 ± 0.068, more than one codon substitution on the average per gene (Ayala *et al.*, 1974*c*). It should be noted that perhaps only one-third of codon substitutions lead to electrophoretically detectable changes in the gene products. Therefore, the genetic differences between populations and species are greatly underestimated.

On the other hand, there is no assurance that the structural genes whose products can be studied by present electrophoretic techniques are an unbiased sample of all genes. We have no information about the rates of divergence in *Drosophila* in regulatory genes or in duplicated or multiply repeated gene loci. The important work of Wilson and his collaborators (King and Wilson, 1975; and Chapter 13) on the evolution of structural genes versus evolution of gene regulation shows how much remains to be learned about the molecular aspects of species formation and evolution. The genetic distance between *Homo sapiens* and chimpanzee determined by examination of 44 gene loci is only 0.62. This is not significantly greater than the 0.58 obtained by Ayala and collaborators for sibling species of *Drosophila!* Sibling species by definition are barely or not at all distinguishable morphologically; man and chimpanzee are placed in different families, Hominidae and Pongidae, on the basis of morphological evidence. If zoological systematics were erected on behavioral and psychological traits, they might, perhaps, have been placed in different classes or even phyla. The question that suggests itself is this: Could the evolutionary divergence in some groups occur by changes in regulatory genes alone with structural genes remaining constant or nearly so?

SUMMARY

It appears that the speciation processes at the molecular level will prove to be as diversified as they are on the organismic level. To find different kinds of species, one does not even need to compare species ranging all the way from *Drosophila* to man and chimpanzee. The patterns of speciation seem to be rather distinct even within the genus *Drosophila*. Witness the continental species such as those of the *willistoni* and *obscura* groups compared with Hawaiian Island endemics studied by Carson and his colleagues. I am reluctant to subscribe to Carson's view (1973) that some speciation in the island forms has involved development of reproductive isolation without ecological adaptive divergence. This is, of course, a problem for future studies. In general, the working hypothesis which seems to me fruitful is that species are not accidents but adaptive devices through which the living world has deployed itself to master a progressively greater range of environments and ways of living. It is well known that adaptiveness and adaptation occur in the living world in remarkably diverse ways. Hence, species as adaptive devices are not the same everywhere. Let us however not miss the unity underlying the diversity.

SUGGESTED READINGS

The biological-species concept generally accepted by biologists was formulated by the main authors of the modern synthetic theory of evolution. The following works should be consulted: T. Dobzhansky, *Genetics and the Origin of Species,* third ed., Columbia University Press, New York, 1951; E. Mayr, *Systematics and the Origin of Species,* Columbia University Press, New York, 1942; G. G. Simpson, *Principles of Animal Taxonomy,* Columbia University Press, New York, 1961; and G. L. Stebbins, *Variation and Evolution in Plants,* Columbia University Press, New York, 1950.

The so-called phenetic concept of species is authoritatively presented by P. H. A. Sneath and R. R. Sokal, *Numerical Taxonomy,* W. H. Freeman, San Francisco, 1973; see, particularly, chapter 7.

The most extensive study of genetic change in the speciation process based on protein differences is summarized in F. J. Ayala, M. L. Tracey, D. Hedgecock, and R. C. Richmond, Genetic differentiation during the speciation process in *Drosophila, Evolution,* **28,** 576–592, 1974.

CHAPTER 7

GENETIC DIFFERENTIATION DURING SPECIATION

JOHN C. AVISE

Biological evolution consists of two processes: anagenesis or phyletic evolution and cladogenesis or splitting. Anagenetic change is gradual and usually results from increasing adaptation to the environment. A favorable mutation or other genetic change arising in a single individual may spread to all descendants by natural selection. Cladogenesis results in the formation of independent evolutionary lineages. Favorable genetic changes arising in one lineage cannot spread to members of other lineages. Cladogenesis is responsible for the great diversity of the biological world, allowing adaptation to a great variety of ways of life. The most decisive cladogenetic process is speciation.

In outcrossing sexual organisms, species are "groups of interbreeding natural populations that are reproductively isolated from other such groups" (Mayr, 1970). Except for impediments of geographic separation, gene exchange can occur among Mendelian populations of the same species. The speciation process requires the development of reproductive isolation between populations, resulting in independent gene pools. Two related questions concerning speciation interest evolutionists: (1) What ecological and evolutionary conditions promote speciation? (2) What changes in the genetic composition of populations result in reproductive isolation?

For sexually reproducing organisms, isolation by geographic barriers and the concomitant severe restriction of gene exchange is the usual prerequisite to genetic divergence and speciation. Geographically isolated populations accumulate genetic differences as they adapt to their different environments. They may become recognizable as races. Not all races will become species, however, because the process of geographic differentiation is reversible. If the races have not sufficiently diverged while allopatric, they may later converge or fuse through hybridization. On the other hand, allopatric populations may sometimes become sufficiently different genetically, so that if the opportunity for gene exchange ensues again, hybrids will have low fitness. Natural selection would, then, favor the completion of reproductive isolation. How much genetic differentiation is required for the development of reproductive isolation between populations?

This chapter will review the evidence, much of it recent and obtained with biochemical techniques, about the amount of genetic differentiation accompanying the speciation process in sexually outbreeding animals. The information available for plants is reviewed in Chapter 8.

METHODS OF DETERMINING SPECIES DIFFERENCES

Early attempts to quantify genetic differences between species involved studying progeny from hybrid crosses, particularly in plants. (See Chapter 6.) It was frequently observed that variability of assorted traits in F_2 hybrids between closely related species was great and that few offspring fell into the parental classes. For example, F_2 hybrids between the tomatoes *Lycopersicon exculentum* and *L. peruvianum* segregate for a great many character differences with such a wide array of recombinations of parental characters that no two individuals appear alike (Rick and Smith, 1953). Such results indicate that the parental species differ in a large number of genes, each with small effects. Few animal species could be successfully hybridized to produce fertile progeny.

A serious difficulty applies to hybridization studies. Classical Mendelian genetics ascertains the presence of a gene by differences in its allelic forms; segregational patterns of alleles are studied in progenies of parents differing phenotypically in characters determined by the gene. Genes which are identical in two populations cannot be detected. Yet to determine the proportion of genes altered during speciation, both variable and invariant genes must be scored. One must be able to study an unbiased sample of the genome; the genes studied should on the average be no more or less different than the remainder of the genome.

In recent years, several molecular techniques have been developed which allow comparisons of genetic material in different populations. These techniques include nucleotide and amino acid sequencing, DNA–DNA and DNA–RNA hybridization, immunological techniques, and protein electrophoresis. These techniques permit quantification of genetic differences between species based on a sample of the genome selected without a priori knowledge of whether the genes sampled are more or less variable than the rest of the genetic material. For the purpose of comparing closely related species, protein electrophoresis has proved to be the most efficient technique. Electrophoretic techniques are briefly described in Chapter 2.

To a first approximation the overall genetic differentiation between populations may be reasonably estimated by examining a number of proteins. Nonetheless, potential sources of bias must be recognized. One possible source of bias is that electrophoretic methods sample only structural genes which code for soluble enzymes and proteins. Other classes of structural genes, as well as regulatory genes whose products are not translated into proteins, cannot be sampled by electrophoresis. Another difficulty is that not all changes in the DNA will be registered in an altered protein mobility because of (1) the degeneracy of the genetic code, (2) the substitution of amino acids with identical net electric charge, and (3) the finite number of distinguishable band mobilities on a gel. Although we do not know how much genetic variation remains undetected by electrophoretic techniques, the results obtained provide only minimal estimates of genetic differentiation.

Electrophoretic data consist of allele and genotype frequencies determined by individually assaying members of populations. A number of indexes have been developed which summarize this information into a common meter of genetic divergence between populations. According to the method proposed by Nei (1972) the normalized genetic *identity* of genes between two populations at the j *locus* is defined as

$$I_j = \frac{\Sigma x_i y_i}{(\Sigma x_i^2 \Sigma y_i^2)^{\frac{1}{2}}}$$

where x_i and y_i represent the frequencies of the ith allele in populations X and Y, respectively. For all loci in a sample, the overall genetic identity of X and Y is defined as

$$I = \frac{J_{xy}}{(J_x J_y)^{\frac{1}{2}}}$$

where J_x, J_y, and J_{xy} are the arithmetic means over all loci of Σx_i^2, Σy_i^2, and $\Sigma x_i \Sigma y_i$, respectively. The accumulated number of codon substitutions per locus (genetic *distance*) since the time of divergence of two populations is estimated as

$$D = -\ln I$$

Another commonly employed index, S, measures the mean geometric distance between allele frequency vectors over all loci (Rogers, 1972). Estimates of I and S calculated from the same data are fairly similar, although S generally gives lower numerical values than I.

GEOGRAPHIC SPECIATION

Two stages may be recognized in the process of geographic speciation (Ayala *et al.,* 1974*a*). During the first stage, populations become isolated by geographic barriers and accumulate genetic differences. Much of this divergence is the result of adaptation to different environments, but other factors such as genetic drift and founding events may play a role. Partial or even complete reproductive isolation between populations may develop as a by-product of this genetic divergence. During the second stage of speciation, natural selection may hasten direct development of reproductive isolation in the form of prezygotic isolating barriers. This stage begins when genetically differentiated populations regain geographic contact. If reproductive isolation is not yet complete and if the gene pools have sufficiently diverged, matings between individuals of different populations may leave progenies with reduced fitness. Natural selection would then favor genetic variants promoting matings between members of the same population. Reproductive isolation would be perfected.

Two survey strategies have been employed in attempts to determine the proportion of genetic loci altered during the speciation process. A direct strategy involves assaying populations which appear to be in various stages of the speciation process. Such studies permit assessment of the amount of genetic differentiation during the first stage of speciation when allopatric populations develop incipient reproductive isolation, and during the second stage of speciation when reproductive isolation is being completed by natural selection between populations which have regained sympatry. A second survey strategy involves assaying populations belonging to different species. Species' differences represent the sum of genetic differences accumulated subsequent to speciation as well as during the speciation process itself. Hence interest has centered on species which by other criteria appear particularly closely related, such as morphologically similar species (sibling species) and species that can hybridize.

GENETIC CHANGE DURING SPECIATION IN *DROSOPHILA*

The most extensive study of genetic differentiation during geographic speciation has been carried out in the *Drosophila willistoni* complex of populations (Ayala *et al.,* 1974). The *willistoni* group consists of more than a dozen closely related species endemic to the American tropics. Some species, such as *D. nebulosa,* are readily distinguishable morphologically from the rest. Other species are called siblings because they are morphologically nearly indistinguishable. Despite their morphological similarity, sibling species are completely isolated reproductively.

Some species consist of more than one subspecies. *Drosophila willistoni willistoni* and *D. w. quechua* are allopatric, separated by the Andes in Peru. Hybrid males, but not females, of laboratory crosses between *D. w willistoni* and *D. w. quechua* are sterile when the mothers are *D. w. quechua,* but there is no evidence of behavioral or sexual isolation between the subspecies. Another pair of subspecies, *D. equinoxialis caribbensis* and *D. e. equinoxialis,* always yield sterile male but fertile female hybrids in laboratory crosses and show no sexual isolation. These subspecies pairs of *Drosophila* represent populations in the first stage of geographic speciation. They exhibit incipient reproductive isolation in the form of partial hybrid sterility. The subspecies are presently allopatric, but if they were to become sympatric, natural selection would favor the development of more complete reproductive isolation between them.

Another set of populations at a more advanced stage within the speciation process is represented by six semispecies of *D. paulistorum.* Two or more semispecies may coexist sympatrically in many localities, and reproductive isolation is in some cases apparently complete. In addition to partial sterility of hybrids in laboratory crosses, sexual isolation between semispecies is well developed in many cases. Nonetheless, gene flow among semispecies may occasionally occur, particularly via the *Transitional* semispecies. Semispecies of *D. paulistorum* are in the second stage of speciation, when reproductive isolation in the form of sexual isolation is being completed under the influence of natural selection favoring homotypic matings.

Populations at five increasing levels of evolutionary divergence may thus be recognized in the *willistoni* group: (1) geographic populations within a taxon; (2) subspecies, in the first stage of geographic speciation; (3) semispecies, in the second stage of speciation; (4) sibling species, in which speciation is complete but little morphological divergence has accumulated; and (5) nonsibling species, exhibiting morphological differences as well as reproductive isolation. Ayala and coworkers have studied electrophoretic variation in proteins encoded by 36 gene loci in populations of each of several *willistoni* group species. Estimates can be made of levels of genetic differentiation between populations at each of the above five levels of evolutionary divergence (Figure 1 and Table I).

Geographic populations are genetically differentiated very little. At most loci, populations have essentially identical allele frequencies, although at an occasional locus allele frequencies vary considerably between populations. Mean genetic identity across loci in comparisons among local populations of the same species is $\bar{I} = 0.970 \pm 0.006$.

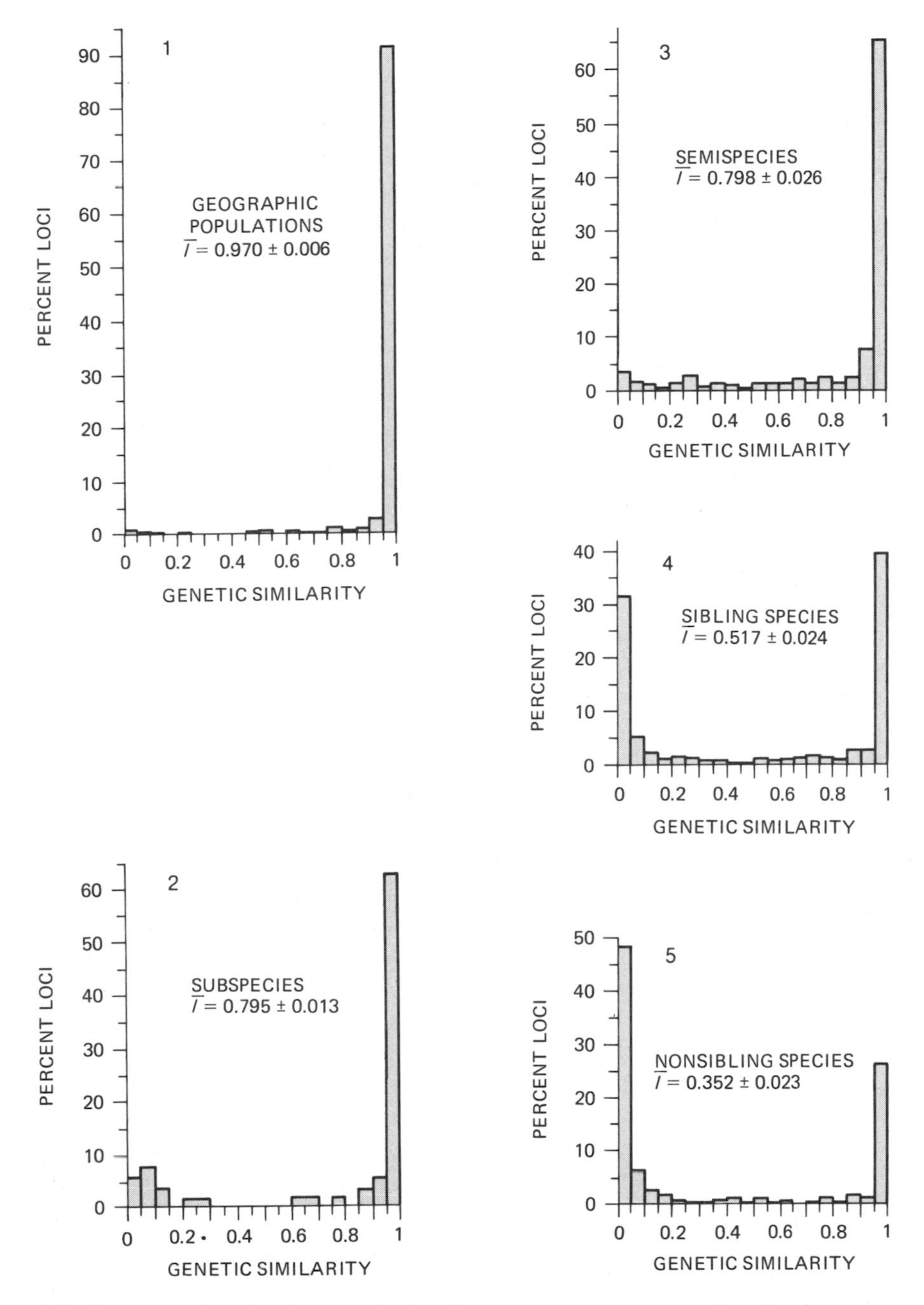

FIGURE 1. Distribution of loci with respect to genetic identity in comparisons among populations in the *Drosophila willistoni* group at five early stages of evolutionary divergence. (After Ayala *et al.*, 1974*b*.)

In contrast, considerable genetic differentiation exists between subspecies. Genetic identities in comparisons between *D. w. willistoni* and *D. w. quechua,* and *D. e. equinoxialis* and *D. e. caribbensis* average 0.808 and 0.782, respectively. Average genetic identity between subspecies in the *willistoni* group equals $\bar{I} = 0.795 \pm 0.013$, and average genetic distance $\bar{D} = 0.230 \pm 0.016$. About 23 allelic substitutions per 100 loci have occurred in the separate evolutions of each pair of subspecies. This genetic divergence represents about an eightfold increase over average distance between local populations. A very substantial amount of genic differentiation has occurred during the first stage of geographic speciation in these *Drosophila* populations.

Mean genetic identities and distances between semispecies of *D. paulistorum* are $\bar{I} = 0.798 \pm 0.026$ and $\bar{D} = 0.226 \pm 0.033$ (Figure 1). These values do not differ significantly from those observed between subspecies. The additional amount of genetic differentiation accumulated during the second stage of speciation when sexual isolation is being developed appears small. It may be that relatively few genes affect mating behaviors and that these may be rapidly selected to provide sexual isolation.

Sibling species of the *willistoni* complex cannot be morphologically distinguished, except by minute differences in male genitalia, yet genetically they are very different from one another: $\bar{I} = 0.517 \pm 0.024$ and $\bar{D} = 0.750 \pm 0.078$. The distribution of loci with respect to genic similarity is strongly bimodal (Figure 1). In the comparison between any two species, most genes are either essentially identical in allelic composition or else completely different. Few genes have I values intermediate between 0.05 and 0.95. The morphological similarity of sibling species belies their large genetic differences.

Morphologically distinguishable species of the *D. willistoni* group differ from one another by an average of more than one electrophoretically detectable allelic substitution per locus: $\bar{I} = 0.352 \pm 0.023$ and $\bar{D} = 1.058 \pm 0.068$. Genetic differences continue to accumulate following the speciation process. By the time *willistoni* group members have evolved sufficiently to be recognizably different morphologically, a large proportion of the genome (nearly 65 per cent) has been almost completely changed in allelic composition.

Several recent reports of genetic differentiation between populations of other *Drosophila* species generally agree with the observations in the *willistoni* group. *Drosophila pseudoobscura bogotana* is an isolated subspecies inhabiting a small territory near Bogota, Colombia. The main subspecies, *D. p. pseudoobscura,* inhabits western North America, Mexico, and Guatemala. Laboratory matings between *D. p. bogotana* and *D. p. pseudoobscura* yield sterile males when the female parent is *D. p. bogotana.* The sterility barriers separating *D. p. bogotana* and *D. p. pseudoobscura* are very similar to those separating *D. w. willistoni* and *D. w. quechua*. Also, *D. p. bogotana and D. p. pseudoobscura* exhibit no detectable sexual isolation. The estimated genetic distance between *D. p. bogotana* and *D. p. pseudoobscura* is $\bar{D}$ = 0.194, a value very similar to that between *willistoni* group subspecies (Ayala and Dobzhansky, 1974).

Zouros (1973) has studied 16 genetic loci in populations of four species in the *D. repleta* group. Little genetic differentiation exists among local populations of the same taxon ($\bar{D}$ = 0.001). Between two *D. mojavensis* subspecies, $\bar{D} = 0.130 \pm 0.011$. These *mojavensis* subspecies exhibit less genetic divergence than the *willistoni* and *pseudoobscura* subspecies and show less reproductive isolation. The *mojavensis* subspecies have different chromosomal inversions and are morphologically distinct, but they produce fertile hybrids in laboratory crosses.

Drosophila mojavensis and *D. arizonensis* are considered separate sibling species, but they exhibit reproductive isolation comparable to that of the *willistoni* group subspecies. In laboratory crosses, all hybrid offspring are fertile except sons of *D. arizonensis* females. Mean genetic identity between *D. mojavensis* and *D. arizonensis* at 16 loci is $\bar{D}$ = 0.243. However, another pair of *repleta* group species, *D. mulleri* and *D. aldrichi*, are fully reproductively isolated (hybrid crosses yield either sterile progeny or no progeny), and yet mean genetic distance is only $\bar{D}$ = 0.124.

These observations illustrate two major difficulties in attempts to formulate general conclusions about genetic differentiation during speciation. First is the tremendous heterogeneity of the biological world. We certainly do not expect all speciations to involve the same amount of genetic differentiation, even when reproductive isolation arises through the process of geographic speciation. Second is the different set of criteria used by taxonomists to recognize subspecies or species differences in different animal groups. By reproductive criteria, *D. mojavensis* and *D. arizonensis* could be considered subspecies. Nonetheless, what is remarkable is the high degree of congruence among results obtained with very different groups of organisms.

Other studies have been conducted in *Drosophila*, as well as in a few other invertebrate groups (review in Ayala, 1975). In general, they support the major conclusions obtained in the *willistoni* group: (1) Very little genetic differentiation exists between local populations within a species; (2) populations showing incipient reproductive isolation, often in the form of hybrid sterility, exhibit significantly greater genetic distances, involving major allelic changes at up to 20 percent or more of structural genes; (3) populations between which reproductive isolation, and therefore speciation, is complete are usually completely distinct in allelic composition at one-third or more genetic loci. These observations strongly support the contention that major genetic changes often accompany the speciation process.

GENETIC DIFFERENCES IN VERTEBRATES

Many studies exist on genetic differentiation between vertebrate populations in early stages of evolutionary divergence. Only a few representative studies will be discussed here. (For an extensive review, see Ayala, 1975.)

Fish

The sunfish genus *Lepomis* contains 11 species, all native to North America. These species are renowned for their ability to hybridize, particularly in disturbed ecological conditions or in the laboratory. F_1 hybrids from 21 different combinations of two species have been found in nature. Nonetheless, the species retain their identity throughout their ranges, even where they are sympatric. Adults of all species can readily be distinguished morphologically. Avise and Smith (1974*a,b*) have studied electrophoretic variation in proteins encoded by 14 genetic loci in 10 *Lepomis* species. The mean genetic distance between *Lepomis* species is $\bar{D} = 0.627 \pm 0.029$ and $\bar{I} = 0.544 \pm 0.016$. Despite their ability to hybridize, *Lepomis* species are very distinct genetically—about 50 percent of loci are completely distinct in allelic composition in a typical comparison between two species. This is about as much genetic differentiation as between sibling species of *Drosophila willistoni.*

Lepomis macrochirus contains two morphologically different subspecies in the southeastern United States. *Lepomis macrochirus macrochirus* and *L. m. purpurescens* are largely allopatric, but they do meet and hybridize in an extensive intergrade zone, which probably represents a secondary contact between allopatrically evolved populations. Little genetic differentiation exists between populations with a subspecies ($\bar{D} = 0.024$). The genetic distance between the subspecies *L. m. macrochirus* and *L. m. purpurescens* is $\bar{D} = 0.171$, a value close to those observed between *Drosophila willistoni* subspecies or semispecies. F_1 hybrids between *L. m. purpurescens* and *L. m. macrochirus* are fertile. Considerable genetic differentiation has preceded the development of reproductive isolation in these subspecies.

Amphibians and reptiles

The salamander genus *Taricha* consists of three species, *T. granulosa, T. rivularis,* and *T. torosa*. Two allopatric subspecies of *T. torosa* are recognized, *T. t. torosa* and *T. t. sierrae.* Hedgecock (1974) and Hedgecock and Ayala (1974) have reported an extensive study of protein variation at about 40 loci in each of these taxa, as well as in a sample of a related species, *Notopthalmus viridescens.* Estimates of genetic similarity at each of four levels of evolutionary divergence are summarized in Table I. These values are of the same magnitude as those for comparable stages of divergence in the *Drosophila willistoni* group. Nonetheless, reproductive isolation between *Taricha* species is of a partly different nature than that in *Drosophila.* There is no detectable postmating isolation in *Taricha;* reproductive isolation is apparently maintained solely by ethological barriers.

Hall and Selander (1973) have studied two chromosomal forms of the iguanid lizard *Sceloporus grammicus* from Mexico. The chromosomal forms (P1 and F6) are polymorphic for centric fissions of chromosomes 1 and 6, respectively, but otherwise have identical karyotypes. The chromosomally differentiated populations are largely allopatric, but they do hybridize and backcross in a contact zone. There is evidence for lowered fitness in hybrids, and the populations could be considered subspecies. There is no evidence for prezygotic isolating barriers. The degree of genetic similarity between F6 and P1 populations (Rogers' coefficient) is $\bar{S} = 0.787 \pm 0.032$, again similar to reports for subspecies of several other organisms.

TABLE I. Estimates of mean levels of genetic similarity at early stages of evolutionary divergence in various invertebrates and vertebrates

Organism	Local populations	Subspecies or semispecies	Species	Genera
Invertebrates				
Drosophila willistoni	0.97	0.79	0.47	
Drosophila obscura	0.99	0.82		
Drosophila repleta	1.00	0.88	0.78	
Vertebrates				
Lepomis (sunfish)	0.98	0.84	0.54	
Cyprinidae (minnows)	0.99	...	...	0.59
Taricha (salamanders)	0.95	0.84	0.63	0.31
Sceloporus (lizards)	0.89*	0.79*		
Mus (house mice)	0.95*	0.77*		
Dipodomys (kangaroo rats)	0.97*	...	0.61*	0.16*
Peromyscus (mice)	0.95*	...	0.65*	

*Genetic similarities calculated using Rogers' technique (1972). All other values are genetic identities calculated according to the formula for *I* given in the text.

Mammals

Two subspecies of the house mouse, *Mus mus musculus* and *M. m. domesticus,* are largely allopatric in Europe, but they do meet and hybridize along a boundary running east to west through central Denmark. Reproductive isolation appears to be maintained through reduced fitness of backcross progeny, because of disruption of coadapted parental gene complexes, perhaps coupled with environmental differences favoring different alleles on either side of the hybrid zone (Selander *et al.,* 1969; Hunt and Selander, 1973). The mean genetic similarity (Rogers' coefficient) between geographic populations of a subspecies is $\bar{S} = 0.90 \pm 0.01$, and between populations of different subspecies it is 0.769

Extensive analyses of protein variation have been conducted between 11 species of kangaroo rats, *Dipodomys* (Johnson and Selander, 1971), and between 16 species of white-footed mice, *Peromyscus* (review in Avise *et al.,* 1974*a,b*). Results of both studies have been summarized using Rogers' coefficient. Little genetic differentiation exists between most geographic populations of the species extensively surveyed; $\bar{S}$ values between any pair of populations are usually greater than 0.95. Mean genetic similarity between species of *Dipodomys* is $\bar{S} = 0.609 \pm 0.187$ and between species of *Peromyscus* is $\bar{S} = 0.648 \pm 0.017$. In these genera, the range of similarities in pairwise species comparisons is large, and values are strongly correlated with other measures of evolutionary divergence, particularly morphology and karyotype. Mean levels of genic differentiation in these groups are roughly similar to those between sibling species in the *willistoni* group and are generally consistent with the thesis that a considerable amount of genetic divergence may have accompanied speciation. Nonetheless, in both *Dipodomys* and *Peromyscus* several species pairs appear to be separated by fewer than 15 to 20 allelic substitutions per 100 loci.

GENETIC CHANGE AND RATE OF SPECIATION

Avise and Ayala (1975*b*) have recently taken a new conceptual approach to determining the relationship between rates of speciation and amount of genetic divergence. The rationale is simple: If two closely related phylads of equal evolutionary age have had different numbers of speciation events in their history and if speciation events are associated with substantial genetic differentiation, then phylads with greater rates of speciation should show greater amounts of genetic differentiation.

Consider two alternative models fully developed by Avise and Ayala (1975*b*): (1) Genetic change in a lineage is a function of time, unrelated to the number of cladogenetic events, and (2) genetic differentiation is proportional to the number of speciation events in the group. Define

$t_i \equiv$ number of time units since the origin of a phylad, $m_i \equiv$ number of time units between cladogenetic events, assumed constant for a given phylad, $k_i \equiv t_i/m_i$ and $l_i \equiv$ number of species generated per cladogenetic event, $l_i > 1$.

Model 1. We first want to assume that genetic changes occur at a constant rate, a, per unit time. Hence the genetic distance (d_i) between two extant species separated t_i time units ago is simply $d_i = at_i$. Assume that no parallel or convergent genetic changes occur in different lineages and that no species becomes extinct. The average genetic distance $\bar{D}$ among all extant species of a phylad is the sum of genetic distances in all pairwise comparisons between extant species divided by the number of pairwise comparisons. We may drop the proportionality constant a and calculate distances in time units. As shown by Avise and Ayala (1975*b*),

$$\bar{D} = m\left(k - \frac{1}{l-1} + \frac{k}{l^k - 1}\right)$$

(subscripts are dropped for simplicity).

Consider two phylads, R and P, of equal evolutionary age and equal l_i's, with R characterized by a higher rate of cladogenesis ($m_R < m_P$). From the above equation

$$\bar{D}_R = t - \frac{m_R}{l-1} + \frac{t}{l^{t/m_R} - 1}$$

$$\bar{D}_P = t - \frac{m_P}{l-1} + \frac{t}{l^{t/m_P} - 1}$$

If t is large relative to m and l, mean genetic distance approximates t in either phylad, and $D_R/D_P \approx 1$.

Model 2. Assume that the genetic distance between any two contemporaneous species is proportional to the number of cladogenetic events, c_i, separating them ($d_i = bc_i$). If we proceed as before and drop the proportionality constant b, it can be shown that

$$\bar{D} = 2k - \frac{l+1}{l-1} + \frac{2k}{l^k - 1}$$

Consider again two phylads, one species-rich (R) and the other species-poor (P), as in model 1. When t is large relative to m,

$$\bar{D} \approx \frac{2t}{m_i} - \frac{l+1}{l-1}$$

and D_R/D_P will be close to m_P/m_R. In other words, the ratio of mean genetic distances in two phylads is approximately the reciprocal of the time units between cladogenetic events in the two phylads.

Of course the models advanced are oversimplifications. Genetic divergence among members of a phylad will be affected by other processes in addition to the number of speciation events and amount of elapsed time. Nonetheless, the models provide qualitatively distinct predictions which may elucidate main effects. When genetic distance between species is a function of time, mean genetic distances in speciose (species-rich) and depauperate (species-poor) phylads of equal evolutionary age are very similar. On the contrary, when genetic distance is a function of the number of speciations in the history of a phylad, the ratio of mean genetic distance separating species in speciose versus depauperate phylads is greater than 1 and increases rapidly as the frequency of speciations in one group relative to the other increases.

Few data exist at present to test the models. Avise and Ayala (1975*a*) have examined genetic differentiation at 24 loci in species belonging to nine genera of minnows (family Cyprinidae) inhabiting California. The family Cyprinidae is highly speciose, containing about 250 species native to North America. Most North American minnows belong to the subfamily Leuciscinae, which first appears in the fossil record of North America in the Miocene. As a group, the California minnows may be fairly representative of levels of evolutionary divergence among all North American Leuciscinae. The average genetic distance among the nine species of minnows examined is $\bar{D} = 0.568 \pm 0.052$.

In contrast to Leuciscinae, the sunfish genus *Lepomis* is very depauperate, containing only 11 living species. Yet the first fossil remains of *Lepomis* are found at the Miocene-Pliocene boundary, and *Lepomis* appears never to have contained a great many species. Mean genetic distance at 14 loci between 10 of the 11 species of *Lepomis* is $\bar{D} = 0.627 \pm 0.029$. The ratio of mean genetic divergence in the speciose minnows versus the depauperate *Lepomis* is 0.91, in good agreement with predictions of model 1 but in conflict with model 2. This result appears inconsistent with the thesis that speciations in these groups of fish have been accompanied by substantial amounts of genetic divergence.

Nonetheless, no definitive conclusions can be drawn from the present data. There are serious questions about the actual evolutionary ages of Leuciscinae and *Lepomis,* about the true number of cladogenetic events in their histories, and about the adequacy of the nine California species as an unbiased sample of divergence among the North American Leuciscinae. Nevertheless, the models summarized in this section should provide a valuable approach in future attempts to determine the influence of speciation on rates of genetic differentiation.

OTHER MODES OF SPECIATION

Among sexually reproducing outcrossing animals, geographic speciation is likely to be the most common mode of speciation. However, speciations may also occur through rapid, initially nonadaptive means. Such *saltational* speciation events may occur through polyploidization, rapid chromosomal reorganizations, changes in breeding system, or other processes. In such cases, speciation may be completed with little or no change at the genic level. For example, new autopolyploid or allopolyploid species contain no new genic material not already present in their ancestors. Polyploidization and changes in breeding system are rare among higher animals, but these events do play important roles in speciation of many plants. (See Chapter 8.)

One mode of saltational speciation, rapid chromosomal reorganization, may be more common among higher animals. Chromosomal number differences resulting from Robertsonian fusions and fissions distinguish populations within each of two rodent complexes, *Spalax ehrenbergi* (Nevo and Shaw, 1972) and *Thomomys talpoides* (Nevo *et al.*, 1974). Pairs of four chromosomal forms of *Spalax* ($2N = 52, 54, 58$, and 60) meet along narrow contact zones in Israel, but they largely remain distinct. Hybridization occurs more frequently between populations differing by only two chromosomes. The chromosomal differences presumably provide effective postmating barriers to reproduction, although the barriers are strengthened by ethological preferences. A similar situation obtains for at least six chromosomal forms of *T. talpoides* in the western United States. Chromosomal forms occur contiguously usually with little hybridization. The mean level of genetic identity at 17 loci between the chromosomal forms of *Spalax* is $\bar{I} = 0.978$; at 31 loci genetic identity between chromosomal forms of *Thomomys* is $\bar{I} = 0.925$. Nevo *et al.* (1974) argue that speciations have been relatively recent in these groups, with little time for the accumulation of genic change. When chromosomal rearrangements play a significant role in the development of reproductive isolation, speciation may be accompanied by relatively little change at the genic level.

SIGNIFICANCE OF STRUCTURAL GENE CHANGES DURING SPECIATION

Estimates of genetic differentiation during speciation are surprisingly uniform. If one considers the grossly different processes which may be involved in the speciations of organisms as different as flies, salamanders, and mammals, the range of genetic identity estimates among subspecies, semispecies, and closely related species is remarkably small. When compared with levels of genetic differentiation among local populations, these estimates conclusively demonstrate that a substantial proportion of genes may be changed in allelic composition concomitant to the speciation process. Typically, about 20 electrophoretically detectable allelic substitutions per 100 loci accumulate before reproductive isolation is completed. Arguments that speciation is normally accompanied by little genic change are clearly refuted.

Estimates of genetic divergence between closely related species support contentions of major genic changes during speciation. Even species which appear closely related by other evidence, such as morphological similarity or hybridizing propensity, are often completely distinct in allelic composition at one-fourth to one-half of their loci. Populations continue to accumulate genetic differences following speciation. By the time they have diverged sufficiently to warrant their placement in different genera by conventional systematic criteria, they share alleles at only a small proportion of loci.

Nonetheless, the range in biochemical similarities between species is large, and some species appear little or no more distinct than do local populations within a species. Taken at face value, these instances argue that not all speciations are accompanied by large genic reorganizations. Not all speciation events involve the same amount of genetic change even when reproductive isolation arises according to the conventional model of geographic speciation.

Evidence from the *Drosophila willistoni* complex suggests that a substantial proportion of genes is changed during the first stage of geographic speciation but that few additional changes occur during the second stage when the development of reproductive isolation by natural selection is taking place. Perhaps the number of genes required to develop reproductive isolation *per se* is small. We would then expect that in situations where natural selection favors the development of reproductive isolation (such as between populations differing in chromosomal numbers or arrangements), change in only a few genes could complete speciation. If this is true and if the populations differing in chromosomal content were similar at the single-gene level, the two emerged species would differ little in genic content. This appears to be the case among the mole rats *Spalax* and the gophers *Thomomys,* for example.

This line of reasoning is also consistent with the finding of similar levels of genetic differentiation in species-rich versus species-poor groups of fish of about equal evolutionary age, which suggests that time is a primary determinant of genetic differentiation. During the first stage of speciation, time may be an important predictor of the accumulation of genic differences as populations adapt to their environments. The development of reproductive isolation *per se* may not substantially increase the level of genetic divergence.

This raises the question of the relevance of structural gene divergence to speciation. Although a large number of structural genes changes normally precede the completion of reproductive isolation, it is conceivable that these allelic substitutions are irrelevant to the development of reproductive isolation. Some authors have argued that structural gene evolution is primarily a function of mutation rates to neutral alleles. Neutral alleles confer identical fitnesses upon their bearers and, hence, could not play a role in the development of reproductive isolation. On the other hand, if structural genes do affect fitness, depending upon the genetic and environmental backgrounds in which they occur, they may play an important role in speciation. Perhaps the structural gene differences which accumulate during the first stage of geographic speciation contribute to the loss of hybridizing ability or to the lowered fitness of hybrids in contact zones; in some cases they would be directly responsible for reproductive isolation, and in other cases they would provide the selective pressure for the completion of speciation. Evidence is rapidly accumulating that much of the structural genic variability within and between natural populations is maintained by natural selection (Lewontin, 1974). Structural genes themselves may contribute significantly to adaptive divergence leading to speciation.

Evolutionary change also involves regulatory genes—those genes responsible for patterns of structural gene activation and expression. A number of models have been proposed which endow regulatory genes with a crucial role in evolution. As applied to speciation, it has been argued that unfavorable interactions of alleles at regulatory loci are primarily responsible for disruption of patterns and timing of structural gene expression, resulting in decreased hybrid fitness. Regulatory gene changes could be primarily responsible for the development of reproductive isolation and hence of new species.

Wilson and his colleagues (1974*a,b;* see Chapter 13) have recently suggested that there may be two types of molecular evolution—one involving structural genes, which goes on at a more or less constant rate, and a second for regulatory genes, which are primarily responsible for reproductive incompatibilities and morphological evolution. Their arguments stem from the observation that evolutionary divergence in proteins does not closely parallel morphological divergence and loss of hybridization potential when very different groups such as mammals, birds, and amphibians are compared. Although different rates of regulatory evolution are one possible explanation for such observations, it could be that conspicuous morphological changes and reproductive isolation involve only a small proportion of the genome.

Published electrophoretic surveys deal solely with soluble products of structural genes, frequently those forming parts of central metabolic pathways. We do not have good estimates of what proportion of the genome encodes such proteins, and hence we do not know the extent of bias in our sample of the genome. However, within particular animal groups, such as closely related mammals or fish, protein differences correspond quite closely to levels of morphological and other divergence described by classical systematists (Avise, 1974). To this extent at least, structural genes certainly provide information which is of biological and evolutionary significance. Furthermore, we should remember that there are other classes of structural genes. For example, a certain class of structural genes, those encoding nonhistone proteins associated with chromatin in cell nuclei, may serve important regulatory functions (Stein *et al.,* 1975). Hypotheses about relative roles of structural and regulatory genes clearly define an important area of investigation into the amount and significance of genetic divergence during speciation.

SUGGESTED READINGS

The electrophoretic studies of genetic differentiation during species formation are comprehensively reviewed by F. J. Ayala, Genetic differentiation during the speciation process, in *Evolutionary Biology,* vol. 8, T. Dobzhansky, M. K. Hecht, and W. C. Steere (eds.), Plenum Press, New York, 1–78, 1975, where references to the relevant literature can be found.

J. C. Avise, Systematic value of electrophoretic data, *System. Zool.,* **23,** 465–481, 1974, justifies the proposal that protein studies may have great significance in taxonomy.

The theory to evaluate the main factors determining rates of genetic change by comparing evolutionary differentiation in species-rich and in species-poor phylads is advanced in J. C. Avise and F. J. Ayala, Genetic change and rates of cladogenesis, *Genetics,* **81,** 757–773, 1975.

CHAPTER 8

BIOCHEMICAL CONSEQUENCES OF SPECIATION IN PLANTS

LESLIE D. GOTTLIEB

A significant issue for understanding speciation concerns whether the genetic divergence and differences in adaptation that characterize species evolve before or after their reproductive isolation. Orthodox theory considers that spatially isolated populations acquire reproductive isolation as an important by-product of adaptive divergence, that the process marks a substantial genetic reconstitution, and that it occurs gradually over long periods of time. The theory asserts that at their origin species exhibit distinct adaptive features and differ genetically from their progenitors. However, recent findings suggest that certain insect species may originate rapidly and with very little genetic change or adaptive divergence, for example, *Drosophila* species in Hawaii (Carson, 1971) and parasitic Tephritid flies (Bush, 1969, 1974). Diploid species of annual plants also may originate rapidly; their very close morphological and ecological similarities to their progenitors suggest that only minor genetic changes are involved, a thesis brilliantly argued by Lewis (1962, 1966, 1973). The critical test is to identify a species very soon after its origin and compare its genome and adaptations with those of its progenitor.

The precise identification of progenitor and derivative species is often possible in diploid annual plants in which the cytogenetic analysis can be combined with both morphological and ecological studies. Several examples have been identified in *Clarkia* (Lewis, 1973; Small, 1971; Vasek, 1964, 1968), *Stephanomeria* (Gottlieb, 1973*a*), and *Gaura* (Raven and Gregory, 1972), making it possible for one to initiate analysis of the genetic and biochemical consequences of the events of speciation.

At the outset of such studies, two questions have primary interest: (1) Does the derivative species possess unique alleles and/or display different frequencies for alleles possessed in common with its progenitor? (2) Has the derivative species acquired distinct physiological capabilities, morphological structures, or other adaptations? The first question is most efficiently approached by assaying the electrophoretic mobility of the enzymes specified by a large number of genes. The second question is answered by comparative analysis of growth rates, fecundity, response to various stresses, competitive ability, and other factors using a combination of growth chambers and transplant techniques.

This chapter describes the results of initial studies in a number of diploid species in which the progenitor and derivative species have been unambiguously identified, with the derivative being of recent origin. In addition, the unique biochemical consequences of speciation via the addition of divergent diploid genomes to yield tetraploid species will be explored using the recently evolved tetraploid species of *Tragopogon* as examples. The evidence presented also shows that when speciation at the diploid level involves chromosomal rearrangement in self-compatible or self-pollinating species, duplication of genes may occur, as appears to have been the case in two species of *Clarkia* (Gottlieb, 1974*a,* and unpublished data).

EFFECT OF MODE OF ORIGIN ON GENETIC DIFFERENTIATION

Reproductive isolation may evolve as a by-product of the slow and gradual accumulation of genetic differences in geographically separate populations through the operation of founder effect, drift, and selection for local adaptation. Under these circumstances, a new species is likely to possess new alleles at many gene loci, perhaps new gene linkages, and possibly new patterns of gene regulation. Such an explanation accounts for the patterns of species or race differences that take the form of a more or less concordant series of small changes in morphology, chromosome structure, and ecological tolerances, for example in *Layia* (Clausen, 1951), *Hemizonia angustifolia* (Clausen, 1951), *Gilia ochroleuca* (Grant and Grant, 1960), *Lycopersicon peruvianum* (Rick, 1963), and *Mimulus* (Vickery, 1974). Similar differentiation would occur whenever reproductive isolation is directly selected in situations where populations which have already acquired distinct adaptive complexes hybridize following secondary contact to produce ill-adapted progenies. Selection for the reinforcement of reproductive isolation initially developed as a by-product of divergence appears to be the situation in several species of

Gilia (Grant, 1966) where reproductive barriers are greater among sympatric species than among allopatric ones.

Alternatively, if speciation follows a rapid and abrupt protocol, the genome of the derivative species is constructed in a few generations from relatively unchanged parental chromosome blocks and contains a limited sample from the parental repertoire of allelic variation and linkages. The derivative species is unlikely to display much phenotypic differentiation or adaptive capabilities—either morphological or ecological—outside the range of its parent, and it is usually recognized only through tests of reproductive compatibilities (Lewis, 1973). Well-studied examples in addition to those already cited include *Holocarpha* (Clausen, 1951), *Chaenactis* (Kyhos, 1965), and the *Gilia inconspicua* complex (Grant, 1964). Reproductive isolation in these species is most often postzygotic and primarily involves chromosomal structural rearrangements. Regardless of their genetic contents, the rearrangements cause abnormalities in chromosome pairing at meiosis. This results in unbalanced distribution of the chromosomes to the gametes, reducing the fertility of hybrids between derivative and parental species.

Chromosomal rearrangements are the primary mode of reproductive isolation in annual plant species and characterize species that presumably originated by gradual processes. Cases of rapid origin of species have been distinguished when certain factors are concordant, particularly, evidence for the near extinction of a parental population, gross differences between parental and derivative species in chromosomal structural arrangement, few morphological differences, and the absence of any population with intermediate expressions of the differences between the parent and derivative (Lewis, 1966). However, it now appears that near extinction of the parental population is not necessary (Gottlieb, 1974*c*) and that the amount of chromosomal rearrangement need not be very great (Gottlieb, 1973*a*). Thus, the critical observations to determine whether a species originated by rapid processes are very substantial morphological similarity to an extant progenitor and the absence of closely related populations with intermediate characteristics. It is also important for purposes of recognition that the speciation events have occurred relatively recently so that postspeciational divergence be minimal.

PATHWAYS OF DIPLOID SPECIATION

The possible pathways of annual plant speciation at the diploid level may be specified in terms of the breeding system of progenitor and derivative species. A three-part symmetry organizes the possibilities: self-incompatible to self-incompatible, self-incompatible to self-compatible, and self-compatible to self-compatible. The latter two pathways include the majority of examples in which the positive identification of progenitor has been accomplished; however, it is not known whether they are more common than the first pathway.

Self-incompatible to self-compatible

The genetic and phenotypic characteristics of a self-pollinating species, relatively soon after its origin, have been determined in "Malheurensis," a recently evolved species, known only from a single locality in the sagebrush desert of eastern Oregon (Gottlieb, 1973*a*, 1974*c*). At this site, "Malheurensis" grows interspersed with a population of *Stephanomeria exigua* ssp. *coronaria* (Compositae), a widespread and ecologically diverse taxon, which was its progenitor. Census estimates made in recent years reveal that the progenitor population is about 50 times larger than that of "Malheurenisis," about 35,000 versus 750 plants. In the field the two species are nearly indistinguishable morphologically; however, in a uniform garden significant differences were demonstrated in several quantitative characters. Their reproductive isolation is maintained by three factors: (1) "Malheurensis" is very highly self-pollinating, which restricts the movement of pollen between it and the obligately outcrossing ssp. *coronaria;* (2) a crossability factor(s) reduces seed set from interspecific cross pollinations compared to conspecific ones by about one-half; (3) if F_1 hybrids are produced, several differences in chromosomal structural arrangement, including a reciprocal translocation, reduce their pollen viability to 25 percent. (Conspecific crosses yield fully fertile offspring.)

The genome of "Malheurensis" is a highly limited extraction of that of the sympatric population of ssp. *coronaria* and appears to possess few unique alleles. This conclusion is based on electrophoretic analyses of enzymes specified by 25 structural genes (Gottlieb, 1973*a,* and unpublished data). The study shows that with the exception of a single rare allele all the alleles of "Malheurensis" are present in ssp. *coronaria,* which, however, has many other alleles in addition. Sixty percent of the genes of ssp. *coronaria* are polymorphic, with a total of 45 alleles, but only 12 percent of the genes, with a total of seven alleles, are polymorphic in "Malheurensis" (Table I). In general, the alleles that characterize "Malheurensis" are either fixed or in high frequency in ssp. *coronaria.* The low genetic variation in "Malheurensis" means that all its individuals can be readily sorted into a small number of highly homogeneous groups on the basis of their electrophoretic phenotypes. Even when variation in morphological characters, such as ligule color, are added, the total number of different phenotypes identified is still only several dozen. In contrast, nearly every individual of ssp. *coronaria* is different when electrophoretic variation alone is considered. The high degree of genetic homogeneity in "Malheurensis" provides additional evidence of the effectiveness of its reproductive isolation.

Physiological differences between the species suggest that "Malheurensis" is maladapted relative to its progenitor. The seeds of ssp. *coronaria* have an obligate requirement for freezing, which delays their germination until the temperature rises in April, but the seeds of "Malheurensis" lack this requirement and germinate readily under cool, moist conditions. This might result in substantial seedling death if the seeds germinate in the fall because of the harshly cold winters that characterize their habitat. Another difference is the wide variation in the number of pappus bristles on different seeds of the same individual in "Malheurensis," which suggests a breakdown of this character which is well-canalized in ssp. *coronaria*.

The relative fitness of the species has been directly tested under a series of controlled conditions in growth chambers. The experiment consisted of growing more than 110 individuals of each species in adjacent pots in seven to 11 independent treatments which differed in soil type, day length, and temperature regime, and measuring 39 characters on each individual to describe its rate of growth, fecundity, and morphology. The results were unambiguous. In every treatment, ssp. *coronaria* grew

TABLE I. The number of alleles (frequency $\geq$ 1%) at polymorphic genes in the progenitor population of *Stephanomeria exigua* ssp. *coronaria* and its sympatric derivative species "Malheurensis"

Gene	"Malheurensis"	Ssp. *coronaria*
LAP-1	1	3
LAP-2	1	2
*EST-1A**	1	2
*EST-1B**	1	2–3
EST-2	2	3
EST-3	3	5
EST-4	1	5
GOT-1	1	3
GOT-3	1	2
GDH	1	2
PGM-1	1	2
*APH-1**	1	4
*APH-2**	1	2
*ADH-1**	1	≈ 4
*ADH-2**	2	≈ 4
	19	45–46

*Genetic analysis incomplete.

larger and more rapidly, initiated stem elongation first, flowered first, and, on the average, was 2 times more fecund. This strong performance was achieved even though ssp. *coronaria* starts its growth with smaller cotyledons. Consistent with its greater genetic variability, ssp. *coronaria* exhibited substantially higher within-treatment variation in these experiments than did "Malheurensis." The consistent substantial performance differential suggests that ssp. *coronaria* is both better adapted and more adaptable than "Malheurensis." Dry-weight measurements of the plants growing at the study site in Oregon showed that similar differences occur under natural conditions.

The apparent lack of either unique alleles or distinct adaptations in "Malheurensis" strongly suggests that its reproductive isolation does not serve to protect complexes of adapted genes and that its origin was fortuitous. Presumably, the species originated in a rapid and abrupt series of events initiated by the occurrence of an individual of ssp. *coronaria* which possessed the ability to pollinate itself (most likely caused by a mutation at a self-incompatibility locus). An inbreeding lineage was formed which acquired one or a few chromosomal rearrangements and became reproductively isolated in the midst of its progenitor population. Fluctuations in the size of the progenitor population appear not to have been required for these events (Gottlieb, 1974*c*). From the genetic point of view, "Malheurensis" can be considered a set of nearly homozygous strains of ssp. *coronaria,* but they are reproductively isolated. (They are interfertile among themselves.) Thus, the initial genetic consequence of speciation in this example was the production of a depauperate species with a restricted range of variation and one that is less well adapted than its progenitor even though it manages to survive, albeit with few individuals.

"Malheurensis" provides a model to explain those frequent cases of speciation in diploid annual plants in which a self-pollinating species is derived from a self-incompatible one. The objection might be made that the restricted genetic variability and apparent maladaptations of the derivative species so limit the probability of its long-term persistence that it should not be considered a "good" species. Such a point of view avoids the critical issue by requiring that species status be reserved for organisms which have distinctive adaptive capabilities. It is only from the study of many examples of recently evolved species in which the progenitor and derivative can be precisely identified and compared with a number of probes that the initial properties of new species can be assessed.

If self-pollinating annual plant species begin inauspiciously like "Malheurensis" but somehow persist and eventually expand their range, do they then show evidence of genetic augmentation? This question can be answered by analyzing *Stephanomeria paniculata,* another diploid, highly self-pollinating derivative of ssp. *coronaria* (Gottlieb, 1972) that very likely had a similar mode of origin to that of "Malheurensis."

Stephanomeria paniculata grows as a weed in open sandy places along roadsides from northern California to eastern Washington and adjacent Idaho and is fully allopatric with contemporary populations of ssp. *coronaria*. The same set of enzymes studied in the Oregon population of ssp. *coronaria* and in "Malheurensis" was also assayed in *S. paniculata*. More than 600 individuals were studied, grown from seed collected separately from nearly 200 plants growing in five widely spaced populations throughout the species range. The remarkable result was that all the populations were monomorphic for the same single allele at each of the 25 genes with the exception that in one population a different allele was fixed at a *GOT* locus. Thus, this species appears to have even less electrophoretically detectable variation than "Malheurensis" even though its populations are found over an extensive region. Although several genes identified in *S. paniculata* have a low frequency in ssp. *coronaria,* all are present in the parent species.

The apparent absence of unique alleles at the large assortment of genes assayed suggests that relatively few genetic changes might transform an initially depauperate genome, which characterizes a self-pollinating diploid species at its origin, into a widespread colonizer. Variation in quantitative characters has not yet been studied in *S. paniculata,* and therefore the true extent of its intraspecific variability is not known. It is possible that the species possesses variation in these characters, as has been found in other predominantly self-fertilizing annuals (Allard, 1965). Because its distribution is now north of the range of ssp. *coronaria* and because it is a weed of open temporary habitats, *S. paniculata* presumably possesses some distinctive adaptations, but the changes in its genome need not have been extensive.

Self-compatible to self-compatible

A number of examples of rapid and abrupt speciation at the diploid level have been identified in *Clarkia*. In every case, the derived species is morphologically very similar to its progenitor, occupies an ecologically marginal habitat, is restricted in distribution, and is reproductively isolated by numerous chromosomal structural differences which substantially reduce fertility of hybrids between the derived and the parental species. Lewis (1973, and references therein) has proposed that speciation in these annual plants results from the rapid reorganization of chromosomes promoted by forced inbreeding among the descendants of survivors of a parental population which had been subjected to a sudden, severe environmental stress such as early-season drought.

Clarkia lingulata and its progenitor *C. biloba* have been extensively studied because cytogenetic analysis permitted an unambiguous identification of their phylogenetic sequence (Lewis and Roberts, 1956). *Clarkia lingulata* is known from only two sites in the Merced River Canyon at the southern geographic periphery of the distribution of *C. biloba* in California. In external morphology, the two species differ only in the shape of the flower petal. Chromosomally, they differ by a translocation, several paracentric inversions, and an extra chromosome in *C. lingulata* which is homologous to parts of two chromosomes of *C. biloba,* causing their hybrids to be sterile.

Because it appears that *C. lingulata* originated recently, comparison of its genome and adaptations to those of its parent species provides a critical test of the consequences of speciation in diploid annual plants. The relative fitness of the two species has been examined in natural habitats outside their ranges, in each other's sites, and in a uniform garden (summarized in Lewis, 1973). Either species can occupy the other's sites (in the absence of the other), but *C. biloba* is probably better adapted than *C. lingulata,* even, to the sites of *C. lingulata.* The failure of *C. lingulata* to outperform *C. biloba* in any of the many different kinds of growth conditions tested strongly suggests that its origin was fortuitous.

If the species originated relatively recently in a rapid series of events as postulated, then most of, if not all, its alleles should also be present in its parent. New alleles result from mutations in DNA and are not affected by chromosomal repatterning. Seven different enzyme systems, specified by eight genes, have been examined by gel electrophoresis in both populations of *C. lingulata* and three of *C. biloba* (Gottlieb, 1974*b*). Concordant with Lewis' results, the two species are highly similar. Three genes are monomorphic for the same allele in both species; with the exception of two alleles at the highly polymorphic *EST–1* gene every allele at the five polymorphic genes in *C. lingulata* is also present in *C. biloba.* In general, the frequencies of the shared alleles are similar, suggesting that their genomes are responding in like manner to similar environments. The genetic identity statistic I (see Chapter 7) demonstrates that the two populations of *C. lingulata* are only slightly more similar to each other ($I = 0.905$) than they are to *C. biloba* ($I = 0.880$). One population of *C. lingulata* is actually more similar to two *C. biloba* populations than to its conspecific population.

Even though *C. lingulata* is substantially more variable than "Malheurensis" and does not display a depauperate genome or show evidence of maladaptation, neither its genetic features nor the results of the comparative growth-rate studies indicate that it possesses distinct adaptive complexes.

Self-incompatible to self-incompatible

Very few species have been positively identified that can illustrate the recent direct derivation (without hybridization) of one obligately outcrossing diploid annual species from another. The paucity of examples presumably reflects difficulty of identification rather than lack of occurrence. One good case appears to be *Gaura demareei* (Onagraceae), a local species of southwestern Arkansas. *Gaura demareei* is morphologically very close to *G. longiflora,* a widespread annual species that grows on light prairie soils throughout the central United States (Raven and Gregory, 1972). The primary difference between them is their mode of pollination; the flowers of *G. demareei* open at sunrise and are pollinated by bees, whereas those of *G. longiflora* open at sunset and are pollinated by moths (Raven and Gregory, 1972).

Both species are chromosomally heterozygous for several reciprocal translocations, but at meiosis the chromosomes regularly disjoin in an alternate manner so that there is no reduction in fertility. Cytogenetic studies of interspecific F_1 hybrids indicate that they differ by at least two reciprocal translocations; however, their mean fertility (70 percent) is higher than usually associated with this amount of chromosomal repatterning (Raven, personal communication).

Although the mode of origin of *G. demareei* has not been specified with the degree of certainty that was possible in the examples in *Stephanomeria* and *Clarkia,* it is worthwhile to compare its genetic variation with that of its progenitor. Ten different enzyme systems, specified by 18 genes, have been examined in an average of more than 30 individuals from two populations of *G. demareei* and three of *G. longiflora* (Gottlieb and Pilz, 1975). Twelve genes, coding seven of the enzyme systems, are monomorphic for the same allele in all individuals of both species. The other genes are polymorphic in both species, with the same alleles in high or low frequencies, respectively. Both species display an identical allelic complement with the exception of several low frequency alleles. The electrophoretic study, albeit limited in terms of number of populations sampled, provides no evidence of enzyme differentiation, suggesting that the species have not diverged genetically in any fundamental way. The difference in the time of opening of their flowers is likely to reflect very few genetic changes, even though it has significant biological consequences.

Summary of genetic consequences of diploid speciation

All studies of genetic differentiation demonstrate that shortly after their origin species are still limited genetic versions of their progenitors or possess very few or no unique alleles insofar as their genomes have been assayed. The additional demonstration that neither "Malheurensis" nor *Clarkia lingulata* possesses distinct adaptations and that both are less well adapted to their own habitats then their progenitors is important concordant evidence at the level of the phenotype. The consistency of these results indicates that the speciation process in many diploid annual plants occurs prior to their acquisition of distinct adaptations and is largely fortuitous.

The findings lend no support to the orthodox view that the genetic consequences of speciation are necessarily the reconstitution of the genome of the derivative species. Just the opposite seems to be the case—at least in annual plants. Species at their outset are extracted from the repertoire of phenotypic variation and genetic polymorphism already present in their progenitors. The genetic consequences of rapid speciation reviewed in this chapter cannot, however, be generalized to all annual plants until species that have originated by gradual processes are identified and tested.

The hypothesis that speciation requires genetic reconstitution (Mayr, 1963) is rejected if the differences between progenitor and derivative are not substantially greater or different in kind than what is observed within species. Clearly it is necessary to formulate operational criteria that can be used to determine when a genetic "revolution" has occurred. Presumably, these criteria will ultimately involve search for changes in types of gene interactions and developmental mechanisms. No case reviewed in this chapter yields evidence of genetic reconstitution. Because very few single changes at appropriate points in development can lead to grossly different phenotypes, it appears unnecessary to construe the differences among closely related annual plant species as being "revolutionary" in any fundamental way.

GENE DUPLICATION IN DIPLOID PLANTS

Electrophoretic analysis distinguishes homozygous and heterozygous enzyme phenotypes and thereby has facilitated the recognition of two putative cases of duplicated genes in the genomes of two diploid species of *Clarkia*. (Absolute proof of duplication requires comparison of amino acid sequences of the polypeptides coded by the duplicated genes.) The initial observation was that all individuals of *C. franciscana* have an identical three-banded phenotype for the dimeric enzyme alcohol dehydrogenase (ADH) but that similar multibanded phenotypes occur only in individuals of the closely related *C. rubicunda* and *C. amoena,* which are heterozygous at a single gene specifying ADH subunits (Gottlieb, 1974*a*). Thus, the ADH phenotype of *C. franciscana* appears to be specified by two genes, each homozygous for an allele that codes a slightly different subunit, and its three ADH variants result from their dimeric associations. The test was to make interspecific F_1 hybrids between *C. franciscana* and individuals of *C. amoena* which were homozygous at a single gene for an allele coding a slow ADH variant. Figure 1 shows that the hybrids displayed a five-banded pattern resulting

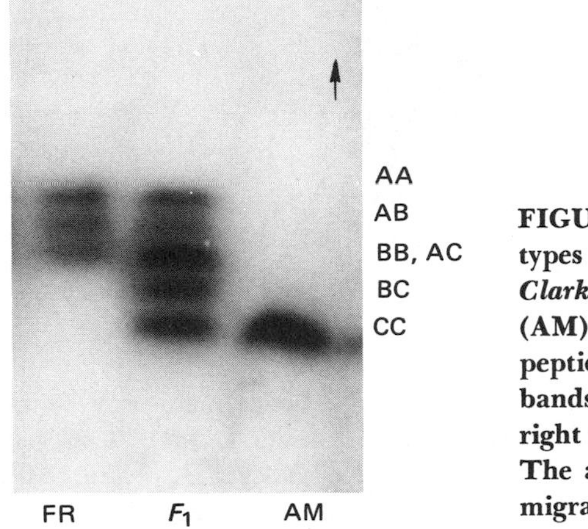

FIGURE 1. Electrophoretic phenotypes of alcohol dehydrogenase in *Clarkia franciscana* (FR), *C. amoena* (AM) and their F_1 hybrid. The polypeptide components of the five isozyme bands of the F_1 are designated on the right in the order of their migration. The arrow shows the direction of migration to the anode.

from the dimeric associations of three different polypeptides; they must have contained three genes.

The highly self-pollinating *C. franciscana* is morphologically very uniform. Electrophoretic study has revealed only a single polymorphic gene out of a total of 12 to 14 genes examined (Gottlieb, 1973*b*). Consequently, the ADH duplication may be particularly felicitous because it provides the species with phenotypic variability characteristic of genetic heterozygotes even though it is very highly homozygous.

A second case of apparent gene duplication has been found in *C. xantiana*. This species has three genes specifying phosphoglucoisomerases (PGI), while only two genes specify them in five other species of *Clarkia*. The alleles at each of the duplicated genes code *PGI* subunits that associate to form both intragenic and intergenic heteromers (Figure 2), whereas in the other clarkias the polypeptide subunits of the two *PGI* genes do not associate to form intergenic heteromeric enzymes. Three alleles have been identified at both the *PGI–2* and *PGI–3* genes of *C. xantiana;* so the duplicated genes do not result in the production of a fixed heterozygote, as is the case with *C. franciscana*. Structural gene duplication in other diploid plants has also been reported for maize *ADH* (Schwartz and Endo, 1966) and maize *EST–5* (MacDonald and Brewbaker, 1974).

If we consider that only about eight to 10 enzyme systems have been investigated in six species of *Clarkia,* the observation of two cases of presumed structural gene duplication suggests that the genetic system of *Clarkia* expedites their accumulation. Two features of the genus make this likely: (1) The origin of species of *Clarkia* is intricately bound up with complex and substantial chromosomal rearrangements; so even

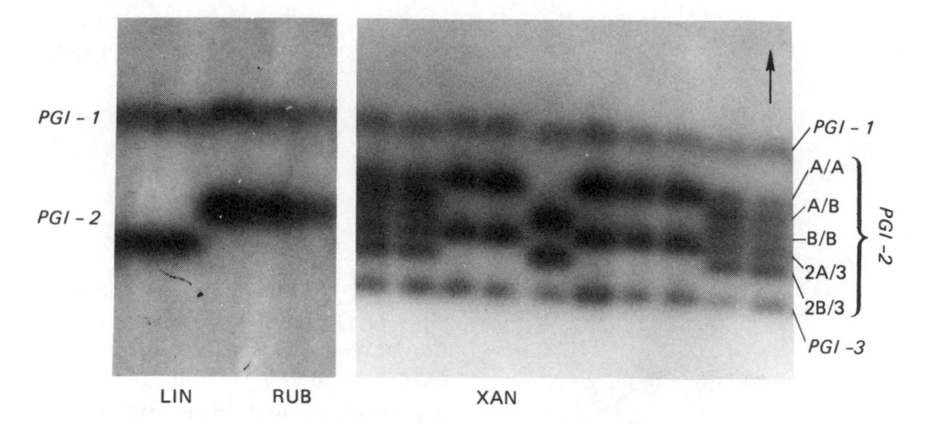

FIGURE 2. Electrophoretic phenotypes of phosphoglucoisomerases in *Clarkia lingulata* (LIN), *C. rubicunda* (RUB), and *C. xantiana* (XAN). Two genes specify PGI in the former two species, but three genes are present in *C. xantiana*. *PGI-2* and *PGI-3* code subunits that associate to form both intragenic and intergenic heteromeric enzymes. The individuals of *C. lingulata* and *C. rubicunda* shown here are homozygous at *PGI-2;* those shown of *C. xantiana* are part of a progeny segregating for two alleles at this gene. Their polypeptide components are designated on the right. The arrow shows the direction of migration to the anode.

closely related species differ by multiple translocations and, occasionally, paracentric inversions. (2) All species of *Clarkia* are self-compatible, which permits their self-fertilization. Duplicated chromosomal segments can be generated in certain progeny classes of crosses between individuals differing for chromosomal rearrangements, particularly insertional translocations, quasiterminal rearrangements, and partially overlapping reciprocal translocations (reviewed in Burnham, 1962; Perkins, 1974). Self-fertilization of the progeny with duplicated segments makes them homozygous in only a few generations. Therefore, *Clarkia* can accumulate duplicated segments, and it is likely that the *ADH* genes in *C. franciscana* and the *PGI* genes in *C. xantiana* mark such segments.

Present hypotheses suggest that self-fertilization is advantageous because it increases the certainty of fertilization (Stebbins, 1957; Lloyd, 1965), perpetuates successful uniform genotypes over generations (Stebbins, 1957), or prevents the disruptive effects of gene flow (Antonovics, 1968). However, the importance of self-fertilization as a means for rapidly making duplicated chromosome segments homozygous suggests that it permits the acquisition of certain biochemical advantages. Rapid increases in fitness may result if the duplicated segments contain genes specifying enzymes or other products which are beneficial either in increased dose or in alternative forms (when duplicated genes contain different alleles). Duplication also leads to the possibility of divergence between polypeptides permitting their differential specialization (Ohno, 1970). Additional advantages are discussed in the next section.

Thus, in many cases self-fertilization might not be directly selected but rather come to characterize plant species because true breeding permits them to acquire certain biochemical advantages. Self-fertilizing species are likely to be perpetuated not as suggested by present hypotheses but because their duplicated genes provide them with increased fitness. It is likely that the duplications originate in association with speciation events. They may not be rare in diploid annual plants because chromosomal rearrangements are a major component of their reproductive isolation and because self-compatibility, with frequent self-pollination, is widespread.

ENZYME MULTIPLICITY IN POLYPLOID PLANTS

Electrophoretic studies have demonstrated that tetraploid plants generally express enzymes of both diploid parents. This has been reported in wheat (Hart, 1969; Mitra and Bhatia, 1971), *Triticale* (Barber, 1970), cotton (Cherry *et al.,* 1970), *Nicotiana* (Smith *et al.,* 1970; Reddy and Garber, 1971; Sheen, 1972), safflower (Efron *et al.,* 1973), *Phaseolus* (Garber, 1974), and *Stephanomeria* (Gottlieb, 1973*c*). If the duplicated genes of polyploids specify different polypeptide subunits of multimeric enzymes, novel heteromeric, or hybrid, enzymes are produced. When the diploid parents differ in the allelic contents at these genes, as is often the case, the tetraploid derivative exhibits novel heteromers that are never produced in either parental species. Such enzymes may have unique biochemical properties (Fincham, 1972; Scandalios, 1974; also see Chapter 3).

In many such cases, the tetraploid species is a "fixed heterozygote" because all individuals express the multiple enzyme phenotypes. Chromosome doubling gives rise in the tetraploid to two pairs of identical homologues that preferentially pair with each other at meiosis. Alleles of each duplicated gene go to the same pole; so each gamete receives one copy of each. Even though each gene is homozygous, a heterozygous phenotype is reconstituted at fertilization.

The multiplicity of enzymes, of which a high proportion are novel and expressed in all individuals of the species, may extend the range of environments in which normal development can take place. Enzyme multiplicity provides a reasonable hypothesis to account for the wider distribution of tetraploid species relative to the diploid progenitors (Fincham, 1969; Barber, 1970; Manwell and Baker, 1970). This may be true even if the tetraploid as a species is less variable than the diploids.

A tetraploid population can maintain a higher proportion of individuals heterozygous at a gene locus than can a diploid population, and individual tetraploid plants have the potential to express more enzymes than diploid ones.

Until recently, comparisons between diploids and tetraploids were often restricted to morphological characters. The general finding was that allotetraploids frequently displayed "gigas" effects (increase in cell size), a changed shape and texture of leaves and flower petals, and intermediacy for polygenically controlled morphological characters (Stebbins, 1971). Many biochemical studies initially emphasized phylogenetic considerations and stressed the additivity of the tetraploid biochemical profile, for example, flavonoids (Smith and Levin, 1963; Brehm and Ownbey, 1965; Levin, 1968), seed proteins (Johnson and Hall, 1965; Cherry *et al.,* 1970; Levin and Schaal, 1970), or increased amounts of diploid constituents (Rick and Butler, 1956).

These results have not clarified one of the most intriguing features of allopolyploidy, namely, that in many plant groups alloplyploids are more widely distributed over more habitats than their diploid progenitors. This is a problem of the first rank because at least one-third of angiosperms and a higher proportion of ferns are polyploid.

The significant problem now is to measure the consequences of the possession of multiple enzymes and demonstrate that they increase the fitness of tetraploid individuals relative to diploids. The possible functions of heterozygosity for structural genes have been investigated from several different standpoints. They may provide heterotic effects and/or novel properties (Schwartz and Laughner, 1969; Schwartz, 1973; Fincham, 1972). The different allelic forms of the same enzyme may possess different biochemical properties which permit more effective metabolism of heterogeneous substrates (Gillespie and Kojima, 1968). Alternatively, additional allozymes may permit plants to maintain a sufficient flux through metabolic pathways in the face of changed reaction conditions associated with varying environments (Chapter 3). The increase in gene number may also result in increased concentration and activity of enzyme products, increasing the flux through rate-limiting reactions; this in turn may lead to an increase in the availability of substrate for other biosynthetic pathways. Enzyme activity in trisomic lines of *Datura* is proportional to the number of coding structural genes present in the genome (Carlson, 1972).

Other studies have suggested that tetraploids may be blocked in specific enzymatic steps, for example, in flavonoid biosynthesis, so that they accumulate compounds that in the diploids are converted to other end products (Levy and Levin, 1971, 1974). In other cases, tetraploids synthesize novel enzymes and convert a compound that is a precursor in the diploids to a novel substance (Levy and Levin, 1971, 1974; Murray and Williams, 1973). These novel enzymes may represent divergent evolution since the origin of the tetraploid, but it is conceivable that their utilization indicates reconstitution of ancestral biosynthetic pathways which had diverged at the diploid level.

The most extensive comparison of enzyme variation in diploid species and their tetraploid derivatives has been done in *Tragopogon* (Compositae) (Roose and Gottlieb, 1975). The situation is unique because the tetraploid species of *Tragopogon, T. mirus* and *T. miscellus,* constitute the only definite examples of the very recent natural origin of polyploid species. Therefore, their genomes can be considered not to have undergone much, if any, modification.

The circumstances of their origin were elegantly described by Ownbey (1950). The three diploid species, *T. dubius, T. porrifolius,* and *T. pratensis,* were introduced to America from Europe during the present century. In southeastern Washington and adjacent Idaho, they became aggressive weeds and have successfully invaded waste places, roadsides, fields, and pastures. The three species are morphologically sharply delimited without overlap in many features. Their populations are frequently sympatric and wherever this occurs, readily detected highly sterile F_1 hybrids are formed. The two tetraploid species originated in this same region, during the last 60 years, following chromosome doubling of interspecific F_1 hybrids. Ownbey (1950) brought them to the attention of science and provided detailed evidence concerning their parentage based on morphological traits, karyotypes, fertility, and genetic analysis: *T. dubius* × *T. porrifolius* gave rise to *T. mirus,* and *T. dubius* × *T. pratensis* to *T. miscellus. Tragopogon miscellus* has become one of the most common weeds in vacant lots in and around Spokane, Washington.

With the generous help of the late Dr. Ownbey, nearly all the populations of both tetraploid species and a number of populations of the three diploid species were collected for electrophoretic studies. The basic evidence sought was the proportion of the tetraploid genome that was fixed in heterozygous state as the result of the combination of different alleles from the diploid parents. The three diploid species are monomorphic for the same allele at nine of the 21 genes examined but are highly differentiated at the other genes. *Tragopogon porrifolius* is the most distinct species because it does not share alleles with either of the other diploid species at six genes and at two other genes is monomorphic for an allele rare or absent in the other species. *Tragopogon dubius* and *T. pratensis* are fixed for alternate alleles at seven genes. Nearly half of the genes distinguish *T. pratensis* and *T. porrifolius.* This substantial allelic divergence among the three diploid species (33 to 48 percent of the genes fixed for alternate alleles depending on the pair of species compared) is concordant with the sharp interspecific differentiation in many morphological characters.

The diploid species possess different alleles at a very high proportion of their genes; hence, their tetraploid derivatives have two alleles and, therefore, are substantially heterozygous. *Tragopogon mirus* expresses an additive pattern for nine genes and *T. miscellus* for seven genes, including five in common with *T. mirus* (Table II). Three of the enzymes showing additive profiles are multimeric, so the tetraploid species express novel heteromeric enzymes with intermediate mobilities in addition to the homomeric ones characteristic of each diploid parent. For example, the alcohol dehydrogenase phenotype of *T. miscellus* contains 13 distinguishable isozymes, five homodimers and eight heterodimers (plus a presumed homodimer and two heterodimers that are not distinguishable because of overlaps on the gel), because its diploid parents have different alleles at each of the three coding genes (Figure 3). Each of the 13 isozymes is fully accounted for by simple additivity of the genes from the diploid parents. Thus, the primary genetic consequence of the combination of the genomes of the diploid species of *Tragopogon* is that the tetraploid species have a fixed heterozygous multienzyme phenotype specified by about 40 percent of the 21 genes examined in each of them. In contrast, fewer than half a dozen individuals in the three diploid species were found to be heterozygous at even a single gene. The substantial heterozygosity in the tetraploids results from their inheriting different alleles from their diploid progenitors which, in turn, reflects the complete divergence between the diploids in

TABLE II. Electrophoretic phenotypes in the tetraploid species of *Tragopogon* illustrating additivity of enzymes specified by alleles inherited from their diploid parents: *T. mirus (T. dubius* × *T. porrifolius)* and *T. miscellus (T. dubius* × *T. pratensis)*. The enzymes are identified by their migration from the origin (in millimeters)

		TRAGOPOGON PHENOTYPES			
Gene	***Porrifolius* →**	***Mirus***	**← *Dubius* →**	***Miscellus***	**← *Pratensis***
EST-2	59	54/59	54	54/57	57
EST-3	53	44/53	44		
EST-4	40	35/40	35, 40		
LAP-1	30	30/32	32	32/35	35
LAP-2	29	25/29	25	25/27	27
APH	52	52/55	55	49/55	49
GDH	16	16/21	21		
G6PD	29	29/31	31		
ADH-1	35	30/35, 35/40	30, 40	30/35	35
ADH-2	...	....	30	18/30	18
ADH-3	...	....	15	5/15	5

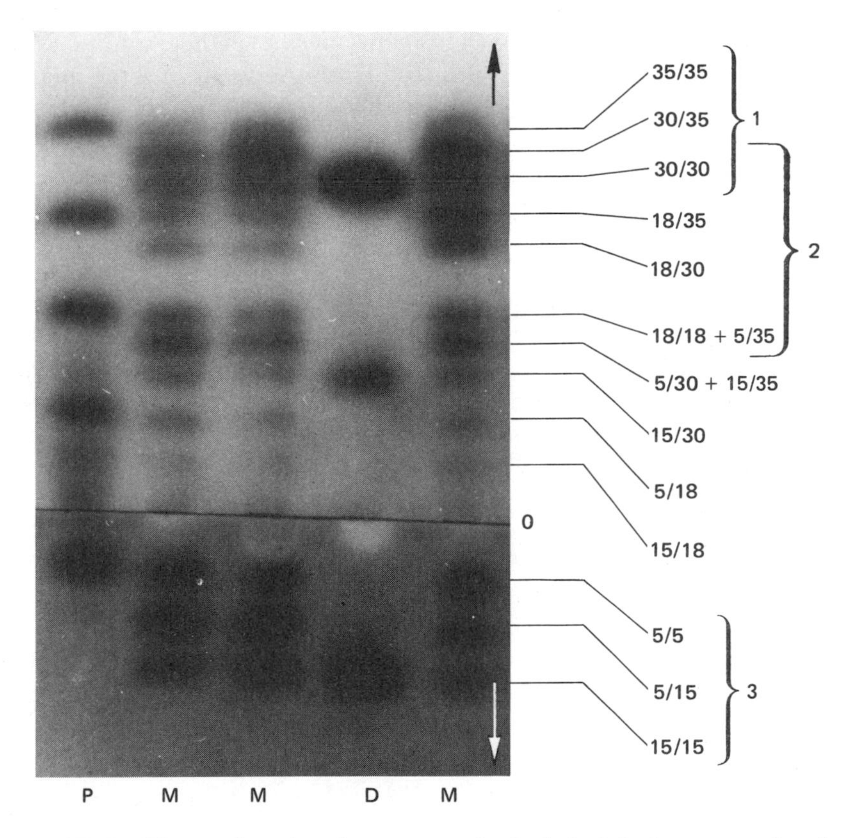

FIGURE 3. Electrophoretic phenotypes of alcohol dehydrogenases in the diploid species *Tragopogon pratensis* (P) and *T. dubius* (D) and their allotetraploid derivative *T. miscellus* (M). Three *ADH* genes are present in both diploid species with polypeptide products that associate to form dimeric enzymes. The 30/30 position in *T. dubius* contains enzymes coded by two genes. *Tragopogon miscellus* forms 15 isozymes as identified on the right by their migration from the origin (in millimeters). The arrows show the direction of migration both to the anode and to the cathode.

allelic constitution at these genes. The proportion of fixed heterozygosity at structural genes in the two tetraploids may actually be higher than presently identified because the electrophoretic technique estimates only a minimum divergence at these genes.

The initial success of a tetraploid species is not a simple function of its amount of enzyme multiplicity. Many other factors are relevant to the likelihood of its establishment, including chromosome pairing behavior affecting fertility, developmental interactions between its genomes, and numerous aspects of its environment. But the demonstration of the biochemical potential of tetraploids in terms of enzyme multiplicity provides the basis for many specific ad hoc studies which were not previously possible. Such studies also serve to link the analysis of evolutionary problems, identified through the study of speciation, with inferences emerging from developmental biology.

SUGGESTED READINGS

Relatively few studies exist on protein differentiation between closely related populations of plants. Besides the works of L. D. Gottlieb reviewed in this chapter, the following should be consulted:

Differences between local populations: M. T. Clegg and R. W. Allard, Patterns of genetic differentiation in the slender wild oat species *Avena barbata, Proc. Nat. Acad. Sci. U.S.A.,* **69,** 1820–1824, 1972; D. A. Levin and W. L. Crepet, Genetic variation in *Lycopodium lucidulum,* a phylogenetic relic, *Evolution,* **27,** 622–632, 1973.

Differences between congeneric species: G. R. Babbel and R. K. Selander, Genetic variability in edaphically restricted and widespread plant species, *Evolution,* **28,** 619–630, 1974.

Two recent important books on plant evolution are V. Grant, *Plant Speciation,* Columbia University Press, New York, 1971; and G. L. Stebbins, *Flowering Plants—Evolution above the Species Level,* Belknap Press, Cambridge, Massachusetts, 1974.

CHAPTER 9

PROTEIN SEQUENCES IN PHYLOGENY

MORRIS GOODMAN

The purpose of this chapter is to indicate how phylogenetic problems can be attacked through the use of protein sequences and how the findings enlarge our understanding of phylogenetic relationships and evolutionary mechanisms. This can be done by focusing attention, as far as species go, on the animal order Primates and, as far as genes go, on globin evolution in the vertebrates. Because of the work of a number of laboratories more is known at the molecular level about Primates than about almost any other order of animal species. Similarly, more is known about the structure, function, and evolution of globins than of almost any other group of proteins.

QUESTIONS OF PRIMATE PHYLOGENY

The issues that might be resolved by investigating primate phylogeny through protein sequences come to light when we consider the way primates are ordinarily classified. Level of anagenetic advance or grade of morphological organization is often the basis for establishing the divisions in the order. Brain size, for instance, increased in certain primate lineages during the last 65 MY, and the magnitude of increase has served as a measure of anagenetic advance. This divides Primates into the suborder Prosimii, containing tree shrews, lemurs, lorises, and tarsiers (the primitive primates with respect to brain size), and Anthropoidea, containing monkeys, apes, and humans (the primates with more advanced brains). Moreover, within the Hominoidea (the most advanced division of the Anthropoidea) the large-sized apes (chimpanzees, gorillas, and orangutans) are treated as a single subfamily, whereas human beings are assigned to a separate family.

Although the suborder Anthropoidea can be considered both a monophyletic taxon and a grade assemblage, the suborder Prosimii is strictly a grade for small-brained primates. Some older evidence suggested a different classification. Tree shrews were not placed in the order; lemurs and lorises were grouped in a taxon called Strepsirhini; and *Tarsius,* which shares certain features with monkeys, apes, and humans, such as a hemochorial placenta, was grouped with them in a taxon called Haplorhini. Many controversies have occurred concerning the position of man in the Primates. Wood-Jones had us closer to *Tarsius,* or at least to the common ancestor of *Tarsius* and Anthropoidea, than to any living nonhuman simians; i.e., he had our ancestral lineage evolving separately for more than 55 MY. Straus had us closer to old world monkeys, and Adolph Schultz regarded the assignment of chimpanzees, gorillas, and orangutans to the ape subfamily Ponginae and man's assignment to the family Hominidae as describing the actual branching or clad relationships within the Hominoidea. On the other hand, Sherwood Washburn has vigorously advocated the view of Charles Darwin and Thomas Huxley that our lineage descended from an ape ancestor resembling in many respects present-day African apes. As we shall see, the evidence from protein sequences leaves little doubt that chimpanzees and gorillas share a more recent common ancestry with human beings than with orangutans or any other living animals. The protein sequence evidence is of two types: an indirect type derived from immunological comparisons and a direct type obtained from actual amino acid sequence comparisons.

IMMUNOLOGICAL EVIDENCE ON PRIMATE PHYLOGENY

Nuttall (1904) carried out the first extensive immunological comparisons of primate sera. Understandably, he used procedures which we would judge inadequate by present-day standards. Nevertheless, his main finding has been repeatedly confirmed. Almost invariably the cross reactions of antibodies to the proteins of human serum with those of nonhuman primate sera are strongest with African apes and least with prosimians; Asiatic apes tend to react more strongly than old world monkeys and old world monkeys more strongly than new world monkeys. Aside from confirming this earlier finding, modern immunological investigations are providing a great deal of new information on the phylogeny of the Primates.

In my laboratory, evolutionary distances among primates are determined by carrying out the antibody-antigen reactions in agar gel in immunodiffusion plates designed to allow a relatively large concentra-

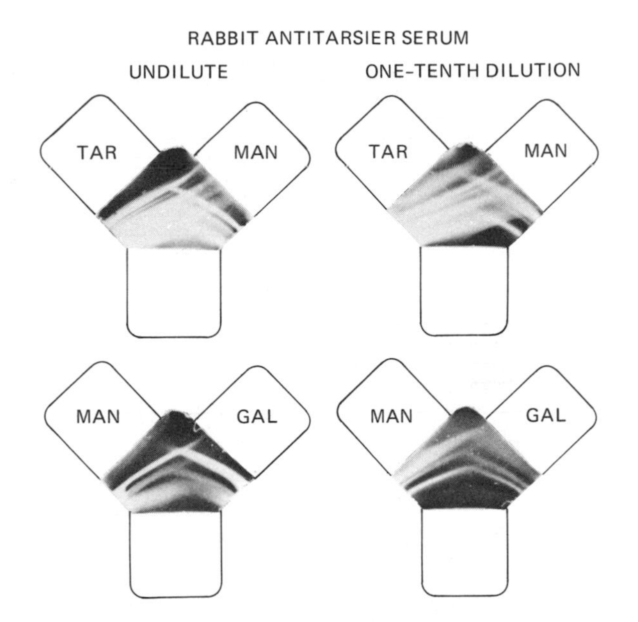

FIGURE 1. The homologous antigen, tarsier serum, develops eight distinct precipitin lines, each of which yields strong spurs against the heterologous antigen, human serum (top two plates). When human serum is compared with another heterologous antigen, galago serum, three precipitin lines develop, with the cross-reacting human proteins yielding positive net spurs against the cross-reacting galago proteins.

tion of reactants to meet in a restricted area of agar and thus to be more sensitive than the usual plates to the smaller divergencies in antigenic structure (Goodman, 1962, 1963*a;* Goodman and Moore, 1971; Goodman *et al.,* 1974*a;* Baba *et al.,* 1975). Antigens of the homologous species are compared with those of other species (the heterologous species), and the heterologous species are also compared with each other. An example of these comparisons, in which *Tarsius syrichta* is the homologous species and *Homo sapiens* and the lorisiform prosimian *Galago crassicaudatus* are heterologous species, is shown in Figure 1.

In a comparison of a homologous antigen against a related heterologous antigen, both antigens form precipitin lines in the agar which grow toward each other and meet on an angle determined by the geometry of the plates. The homologous precipitin line continues to grow even after it is met by the heterologous line, and this growth, called a spur, becomes larger as the heterologous antigen shares fewer antigenic determinants with the homologous antigen. When two heterologous

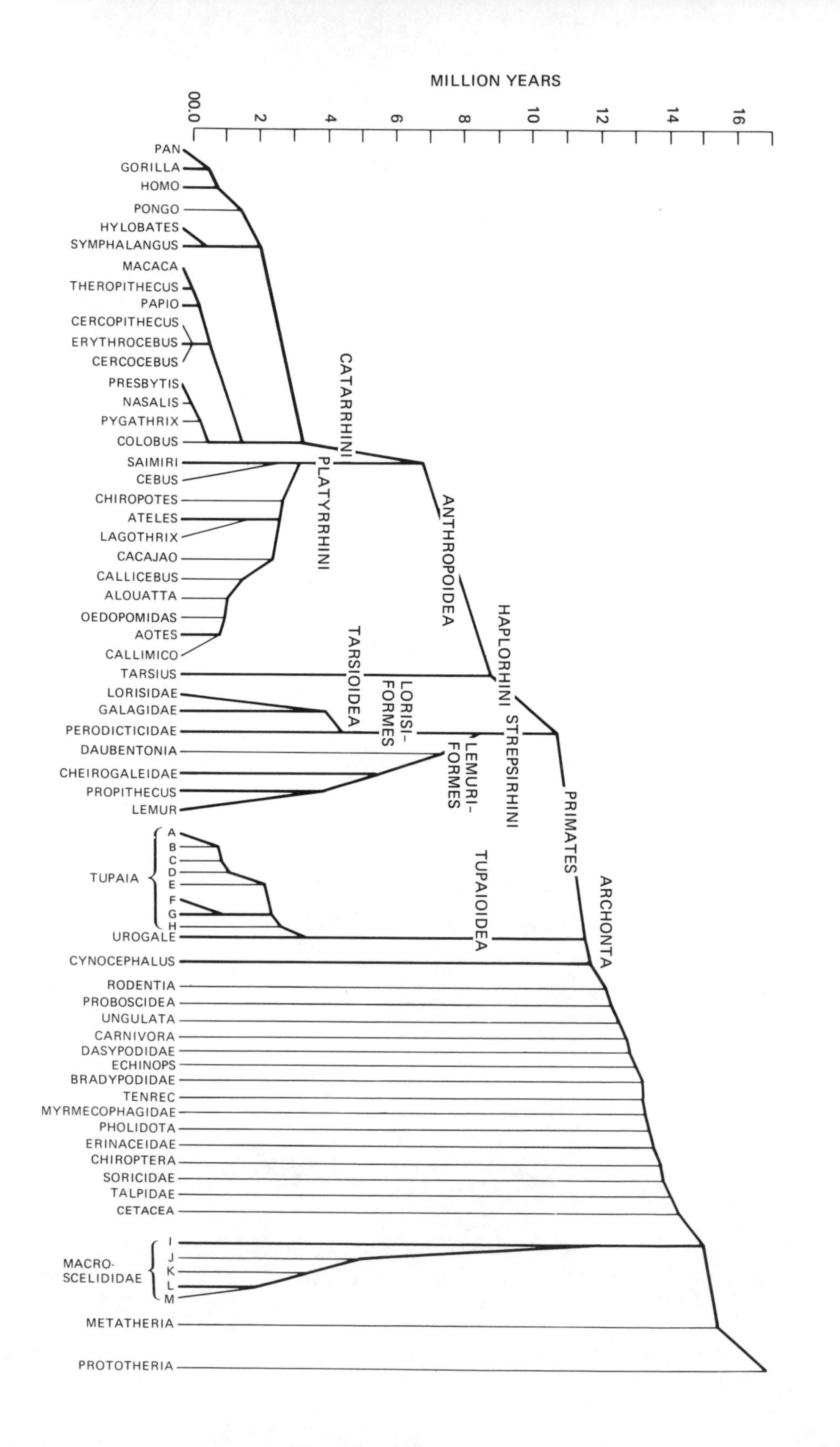

MILLION YEARS
00.0
2
4
6
8
10
12
14
16
PAN
GORILLA
HOMO
PONGO
HYLOBATES
SYMPHALANGUS
MACACA
THEROPITHECUS
PAPIO
CERCOPITHECUS
ERYTHROCEBUS
CERCOCEBUS
PRESBYTIS
NASALIS
PYGATHRIX
COLOBUS
SAIMIRI
CEBUS
CHIROPOTES
ATELES
LAGOTHRIX
CACAJAO
CALLICEBUS
ALOUATTA
OEDOPOMIDAS
AOTES
CALLIMICO
TARSIUS
LORISIDAE
GALAGIDAE
PERODICTICIDAE
DAUBENTONIA
CHEIROGALEIDAE
PROPITHECUS
LEMUR
TUPAIA
A
B
C
D
E
F
G
H
UROGALE
CYNOCEPHALUS
RODENTIA
PROBOSCIDEA
UNGULATA
CARNIVORA
DASYPODIDAE
ECHINOPS
BRADYPODIDAE
TENREC
MYRMECOPHAGIDAE
PHOLIDOTA
ERINACEIDAE
CHIROPTERA
SORICIDAE
TALPIDAE
CETACEA
MACRO-
SCELIDIDAE
I
J
K
L
M
METATHERIA
PROTOTHERIA
CATARRHINI
PLATYRRHINI
ANTHROPOIDEA
HAPLORHINI
TARSIOIDEA
LORISI-
FORMES
LEMURI-
FORMES
STREPSIRHINI
PRIMATES
TUPAIOIDEA
ARCHONTA

antigens are compared, the one closer to the homologous yields the longer spur against the other. The observations are recorded and by a suitable computer analysis (Moore and Goodman, 1968) converted into tables of antigenic distances. From the tables for the different homologous species (those against which antisera were produced) a divergence tree is constructed of the species compared. Species which show the least antigenic divergence are grouped first, whereas the largest and most divergent of the species agglomerations are joined last.

Figure 2 shows an antigenic divergence tree based on over 4,800 immunodiffusion plate comparisons developed with rabbit antisera to serum proteins of 20 primate, three tree shrew, one flying lemur, and two elephant shrew species. *Tarsius* and Anthropoidea are slightly closer to each other than to other primates, and similarly Lorisiformes and Lemuriformes are slightly closer to each other than to other primates. These primate lineages are closer to each other than to tree shrews and flying lemur (*Cynocephalus*) but more like tree shrews and flying lemur than other mammals. In turn tree shrews (Tupaioidea) and flying lemur are about equidistant from each other and from Primates but closer to each other and to Primates than to other mammals. The primate-like affinities of flying lemur in immunological reactions were first observed by Sarich and Cronin (1974).

In the lorisiform branch, Asian lorises (*Loris* and *Nycticebus*) are not closer as traditionally depicted to African pottos (*Perodicticus* and *Arctocebus*) but diverge more from them than from African bushbabies (*Galago* and *Galagoides*). Thus our divergence tree shows three major lorisiform lineages: Lorisidae, Perodicticidae, and Galagidae. Within Lemuriformes, mouse and dwarf lemurs (*Microcebus* and *Cheirogaleus*) are usually treated as a subfamily of Lemuridae, and these small- and medium-sized lemurs are grouped with large-sized ones (*Propithecus*) into Lemuroidea. On the basis of reexamining the morphology of these animals, Szalay and others suggest that the mouse and dwarf lemurs, or cheirogalids, are not members of Lemuridae or even of Lemuroidea but instead are a branch of Lorisiformes. The immuno-

FIGURE 2. A phylogenetic tree based on immunodiffusion plate comparisons. Heavy lines descend to taxa used as homologous species; fine lines descend to taxa used only as heterologous species. The *Tupaia* species are (A) *T. chinensis*, (B) *T. longipes*, (C) *T. glis*, (D) *T. minor*, (E) *T. palawensis*, (F) *T. montana*, (G) *T. tana*, and (H) *T. belangeri*. The Macroscelididea are (I) *Rhynchocyon*, (J) *Petrodromus*, (K) *Nasilio*, (L) *Elephantulus myurus*, and (M) *E. intufi*.

diffusion results indeed remove cheirogalids from Lemuroidea but still keep them as Cheirogaleoidea in Lemuriformes.

Anthropoidea divides into Platyrrhini for new world monkeys, including marmosets, and into Catarrhini for old world monkeys and hominoids. Old world monkeys consist of one extant family, Cercopithecidae, whose genera are divided into Colobinae and Cercopithecinae, exactly as in traditional schemes. Hominoidea splits into Hylobatidae for the gibbons and siamangs and into Hominidae for large-sized apes and humans. Hominidae, in turn, splits into Ponginae for orangutans and Homininae for gorillas, chimpanzees, and humans.

It may be noted that the antigenic distance between the genealogical Ponginae and Homininae is about the same as the distance between Colobinae and Cercopithecinae, the two subfamilies of old world monkeys (1.56 compared with 1.50). In fact, the analysis of King and Wilson (1975) shows that at the protein sequence level the difference between chimpanzees and human beings is no more than that of closely related species in an average mammalian genus. This small degree of sequence divergence is consistent with the proposal that the African apes should be represented as taxonomically closer to humans than to orangutans. However, it is not my reason for suggesting such a classification. My reason is simply the belief that zoological classification should give priority to genealogy. When this is done, there is less room for misinterpretations of what the classification says and no room for the subjective bias entailed on giving taxonomic weight, as traditional classifications do, to considerations concerning an organism's ecological niche and type of adaptive specializations. The relationships determined by genealogy are genetic relationships. They reflect the most stable and fundamental qualities of organic beings and thus provide the real basis for a natural or truly evolutionary system of classification.

The immunological data gained in the laboratories of Allan Wilson and Vincent Sarich by the microcomplement fixation technique with antisera to purified serum albumins and transferrins of primates and other mammals agree with the main course of primate phylogeny depicted in Figure 2 except for the branchings at the origin of the primate radiation. Sarich and Cronin (1974) find Anthropoidea, *Tupaia,* a lorisiform-lemuriform branch, *Tarsius,* and flying lemur to be all equidistant from one another, whereas with respect to a range of serum proteins we find *Tarsius* to be slightly closer to Anthropoidea (Figure 1) but both Tupaioidea and flying lemur to be more distant from Anthropoidea than either *Tarsius* or the lemurs and lorises.

Hemoglobin sequences (next section) and repeated DNA sequences (Hoyer and Roberts, 1967) support our finding that *Tarsius* is slightly closer to the Anthropoidea and that *Tarsius* and the lemurs and lorises are clearly closer than tree shrews. On the other hand, with respect to the protein antigens of lens (Maisel, 1965) *Tarsius* and Lorisiformes are highly similar and diverge from Anthropoidea. Nevertheless Anthropoidea, *Tarsius*, and Lorisiformes share more lens antigen commonalities with one another than with tree shrews.

AMINO ACID SEQUENCE EVIDENCE ON PRIMATE PHYLOGENY

The amino acid compositions and sequences of such protein chains as globins, fibrinopeptides, carbonic anhydrases, and cytochromes *c* have been determined in a variable number of species of primates and other mammals. In phylogenetic studies, sequences of the same protein type in the different species are aligned against one another, and the minimum mutation distance between them is calculated. These distances are used to construct a divergence dendrogram.This dendrogram then serves as a starting point for one's examination of the many alternative branching arrangements by the maximum parsimony or maximum homology method (Moore *et al.*, 1973; Goodman *et al.*, 1974*b;* Moore, 1975). The method reconstructs for a given dendrogram the ancestral sequence that yields the fewest mutations over the tree. The object in examining alternative branching arrangements is to find a true topology which requires the minimal amount of evolutionary change. Such a tree maximizes the homology between ancestral and descendant sequences. The reconstruction of the amino acid and underlying codon sequence ancestors at the branch points of the tree allows the maximum parsimony method to capture superimposed nucleotide replacements and thus a larger proportion of the genetic changes which actually occurred in the descent of the protein sequences. Moreover a parsimony tree does not depend for its phylogenetic or genealogical accuracy on comparable amounts of change occurring on the different evolutionary branches, as is the case with a divergence tree. Evolutionary rates in the parsimony tree can vary markedly from one lineage to another and within a lineage from one sequence position to another. Such variations in rates illuminate, as we shall see, the role of natural selection in protein evolution.

The parsimony genealogies for sequences of alpha hemoglobin chains, beta-type hemoglobin chains, myoglobin chains, fibrinopeptides, carbonic anhydrases I and II, and cytochromes *c* have been analyzed for the information they provide on primate phylogeny (Goodman 1973, 1974). Each genealogy, as illustrated in Figure 3 for that of alpha hemoglobin chain sequences, depicts Anthropoidea as a monophyletic taxon with two major branches, Platyrrhini and Catarrhini, and each has Catarrhini split into Hominoidea and Cercopithecoidea.

All six genera of extant Hominoidea are respresented in the fibrino-

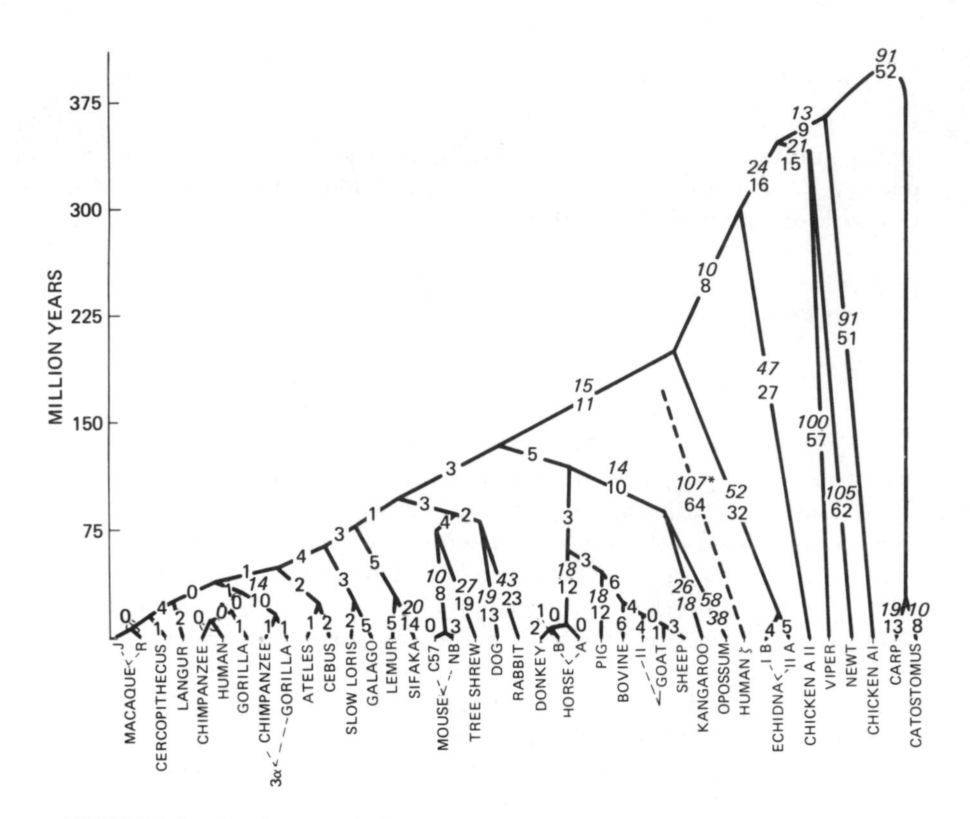

FIGURE 3. Parsimony phylogenetic tree of 39 orthologous and paralogous alpha hemoglobin chain sequences. Inferred as well as rigorously determined sequences are included. Link lengths represent the numbers of nucleotide replacements between adjacent ancestral and descendant sequences; italicized numbers are the link lengths corrected for superimposed replacements by the augmentation algorithm described in Goodman *et al.* (1974*b*). The ordinate scale is inferred from fossil evidence on ancestral splitting times of organisms represented by orthologously related sequences. References to most of these sequences are in Goodman *et al.* (1974*a*, 1975).

peptide tree. As in the immunodiffusion divergence tree, a branch to gibbon and siamang separates from a branch to orangutan, chimpanzee, *Homo,* and gorilla. The orangutan branch separates from a hominid branch that contains identical chimpanzee, human, and gorilla sequences. In the trees for beta hemoglobin and myoglobin chains there are gibbon, human, and African ape sequences (chimpanzee and gorilla in the beta tree and chimpanzee in the myoglobin tree). Again, in each genealogy the African ape and human branches are closer cladistically to each other than to the gibbon branch. In the carbonic anhydrase tree (Tashian *et al.,* 1972), human, chimpanzee, and orangutan sequences represent Hominoidea. Once more, man and chimpanzee are

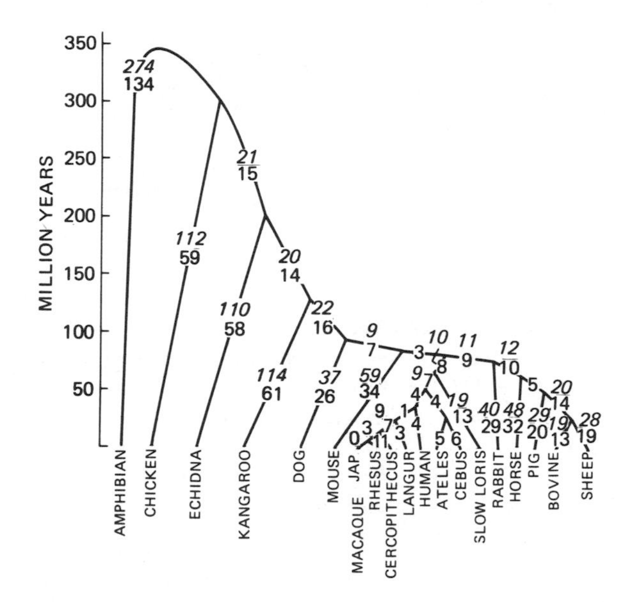

FIGURE 4. Parsimony phylogenetic tree for 19 taxa represented by an extended alignment of operational orthologous alpha and beta hemoglobin chain sequences. For each taxon the sequence of at least one chain, if not both, had been two-thirds or more completed by actual sequencing procedures rather than inferred from amino acid compositions. Augmented link lengths are the italicized numbers. The taxon "Amphibian" consists of newt alpha chain plus frog beta chain.

cladistically closest to each other. Hemoglobin amino acid compositions of orangutan (Buettner-Janusch *et al.,* 1969) and other hominoids and nonrepeated DNA sequences (Hoyer *et al.,* 1972) of hominoids further show that chimpanzees and gorillas diverge less from human beings than from orangutans and gibbons.

Genealogical trees of species have also been constructed by the parsimony method after the sequences of different proteins were combined. Figures 4 and 5 show such genealogies constructed from extended alignments consisting of alpha and beta hemoglobin sequences in Figure 4, and of myoglobin as well as alpha and beta hemoglobin sequences in Figure 5. Again we observe that Anthropoidea splits into Platyrrhini and Catarrhini and Catarrhini divides into Cercopithecoidea and Hominoidea. Lorisiformes (Figures 4 and 5) and Lemuriformes (Figure 5) are closer cladistically to Anthropoidea than to nonprimates and when represented in the same data set (Figure 5) join together first before joining the Anthropoidea. The grouping from immunodiffusion data (Figure 2) of

FIGURE 5. **Parsimony phylogenetic tree of 12 mammalian species represented by an extended alignment of operational orthologous myoglobin and alpha and beta hemoglobin chain sequences. Augmented link lengths are the italicized numbers. The taxon "Ateline" consists of woolly monkey myoglobin and spider monkey alpha and beta sequences. "Lorisid" consists of *Galago* myoglobin and slow loris alpha and beta sequences, and "Lemurine" consists of *Lepilemur* myoglobin and *Lemur* alpha and beta sequences. Except for dog and pig, each species is represented by 440 residue positions (153 + 146 + 141 for myoglobin and the two hemoglobin chain sequences). Because of incomplete myoglobin sequences dog is represented by 409 residue positions (122 + 146 + 141) and pig by 401 residue positions (114 + 146 + 141). Thus, the link lengths of dog and pig are probably slightly less than they would have been if complete myoglobin sequences had been available.**

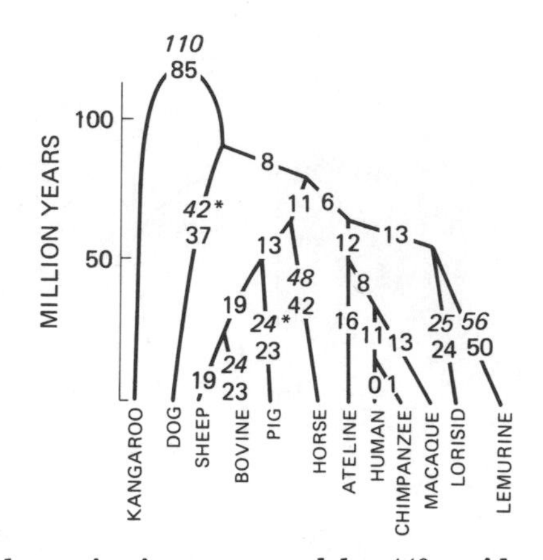

these two prosimian branches into the taxon Strepsirhini is thus supported by sequence evidence. A crucial question which remains concerns the position of *Tarsius* in the Primates. Recently Jan Beard (1975) determined the amino acid sequence of the major hemoglobin from *T. bancanus*. *Tarsius,* slow loris, and Anthropoidea are roughly equidistant with respect to minimum mutation distances calculated in pairwise comparisons of their sequences. However, parsimony analysis shows Anthropoidea and *Tarsius* to be closer cladistically to one another than to Lorisiformes. This further supports the subdivision of the Primates into Haplorhini and Strepsirhini.

In addition to revealing phylogenetic relationships, parsimony analysis provides evidence about rates of molecular evolution. The idea from the neutralist theory of Kimura (1968, 1969) that rates of molecular evolution are generally uniform in different lines of descent is not supported. It can be noted in both the gene and species phylogenies that marked differences in amounts of genetic change during descent occur between lineages. For example in Figure 5, from the common ancestor of eutherian mammals to the present 45 nucleotide replacements can be counted in the human lineage, whereas 83 replacements occur in the lemurine lineage and 75 in the bovine. The amounts of change on the branches in Figure 5 also emphasize the smallness of the genetic difference between man and chimpanzee. Only one nucleotide replacement separates them, yet 43 replacements separate the two bovids, ox and sheep. Such findings highlight the possibility that, on the one hand, Wilson and Sarich (1967, 1969) may be right in arguing that the ancestral human–African ape separation is much more recent than traditional morphological evidence would indicate and, on the other hand, that I too might be right (1961, 1963*b*) in thinking that molecular evolution decelerated in the hominoids. The two ideas are not mutually exclusive! Because my idea is based on an interpretation in which natural selection shapes the long-term patterns of molecular evolution, I shall develop this interpretation by turning now to the second topic—globin evolution during the broad span of vertebrate phylogeny.

AMINO ACID SEQUENCE EVOLUTION OF TETRAMERIC HEMOGLOBIN

The reconstruction by parsimony analysis of the gene phylogeny of globin sequences in the descent of vertebrates has allowed my colleagues and myself to challenge the claims, which gained currency in recent years, that sequence evolution is uniform over time, is dominated by neutral mutations, and is unrelated to organismal evolution. From the patterns of mutations between reconstructed ancestral and descendant sequences of the globin genealogy shown in Figure 6 we have been able to deduce that early evolution of globin sequences in vertebrates proceeded at a faster rate than later evolution. We could attribute the fast evolution to positive natural selection for improvements in the function of hemoglobin and the slow evolution to stabilizing natural selection after the improvements were fixed. By taking advantage of the identification of functional sites in crystalline tetrameric hemoglobins by Perutz and his coworkers (1972 and references therein) we have shown that the sequence sites where the most mutational change occurred during the evolutionary transition from a monomeric hemoglobin to a sophisticated allosteric tetramer were precisely the sites responsible for the advance in functions. Inasmuch as the improvements in function were required in relation to changes in the mode of life of the developing vertebrates, we have concluded that globin sequence evolution is highly correlated with organismal evolution.

The fast evolution in early vertebrates and the slow evolution in later vertebrates is inferred by using a time scale based on paleontological views. For example, the ancestral split between birds and mammals (the two higher branches of amniotic tetrapods) is taken at 300 MY ago, the ancestral split between amphibians and amniotes at 340 MY ago, the teleost-tetrapod split at 400 MY ago, the agnatha-gnathostome (lamprey-jawed vertebrate) split at 500 MY ago, and the mollusc-chordate split at 680 MY ago. About the same scale for vertebrate branch times was used by Kimura and later by Dickerson (1971) when they each

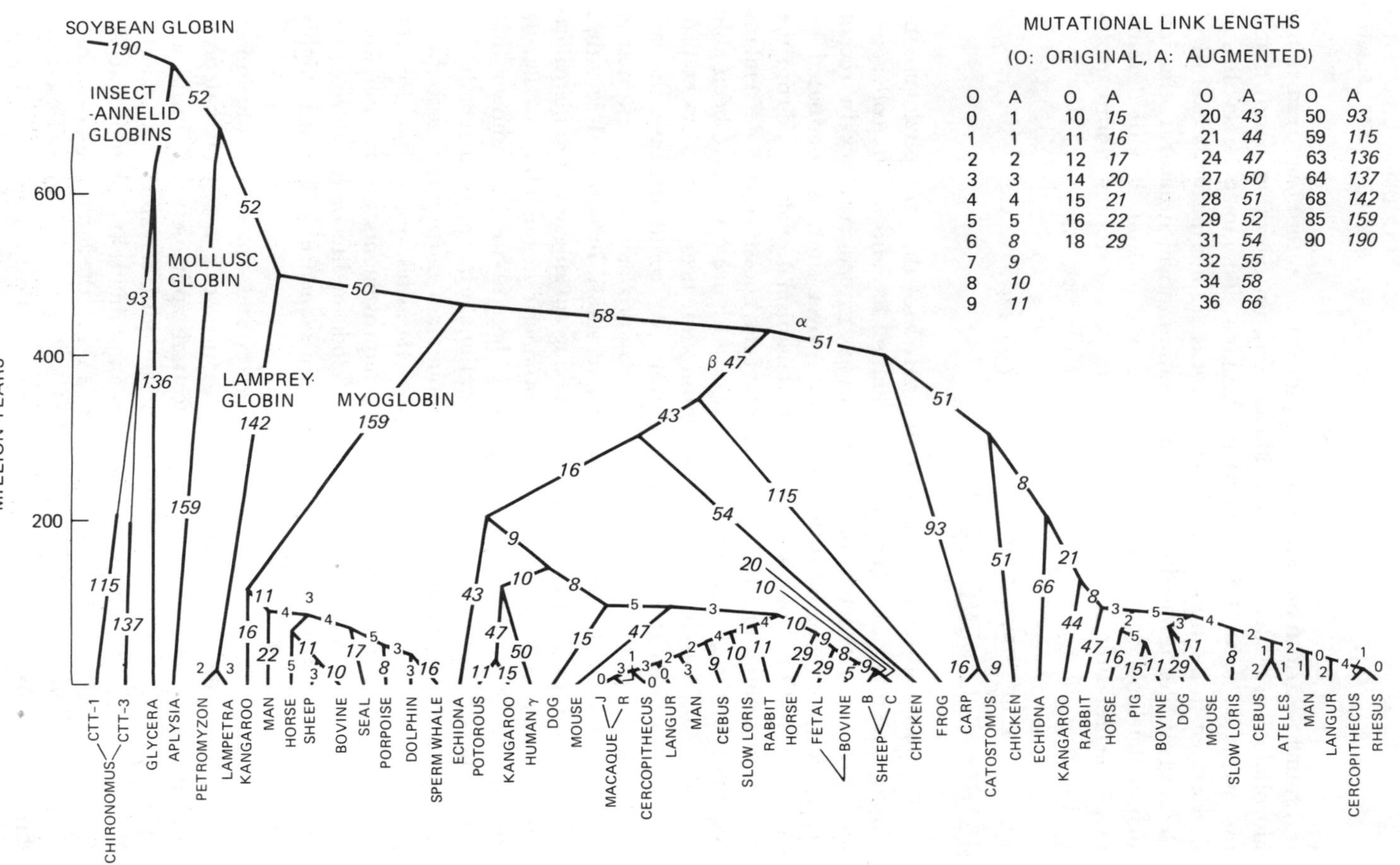
MUTATIONAL LINK LENGTHS
(O: ORIGINAL, A: AUGMENTED)
O A: 0 1, 1 1, 2 2, 3 3, 4 4, 5 5, 6 8, 7 9, 8 10, 9 11
O A: 10 15, 11 16, 12 17, 14 20, 15 21, 16 22, 18 29
O A: 20 43, 21 44, 24 47, 27 50, 28 51, 29 52, 31 54, 32 55, 34 58, 36 66
O A: 50 93, 59 115, 63 136, 64 137, 68 142, 85 159, 90 190
SOYBEAN GLOBIN
INSECT -ANNELID GLOBINS
MOLLUSC GLOBIN
LAMPREY GLOBIN
MYOGLOBIN
α
β
MILLION YEARS
600
400
200
CHIRONOMUS
CTT-1
CTT-3
GLYCERA
APLYSIA
PETROMYZON
LAMPETRA
KANGAROO
MAN
HORSE
SHEEP
BOVINE
SEAL
PORPOISE
DOLPHIN
SPERM WHALE
ECHIDNA
POTOROUS
KANGAROO
HUMAN γ
DOG
MOUSE
MACAQUE J R
CERCOPITHECUS
LANGUR
MAN
CEBUS
SLOW LORIS
RABBIT
HORSE
FETAL BOVINE
SHEEP B C
CHICKEN
FROG
CARP
CATOSTOMUS
CHICKEN
ECHIDNA
KANGAROO
RABBIT
HORSE
PIG
BOVINE
DOG
MOUSE
SLOW LORIS
CEBUS
ATELES
MAN
LANGUR
CERCOPITHECUS
RHESUS

concluded that globins evolved at a uniform rate over geological time. However, they failed to account sufficiently for superimposed mutations in their calculations of distances between contemporary globin sequences; thus they more grossly underestimated the distances between the more anciently separated globins. Our calculations use the numbers shown on the branches in Figure 6; they represent the nucleotide replacement values or mutations separating ancestral and descendant sequences. The evolutionary rates calculated from these values are in nucleotide replacements per 100 codons per 10^8 years (NR%).

For the first 380 MY or so from the invertebrate-vertebrate to the amniote (chicken-mammal) ancestor, the hemoglobin genes evolved at an average rate of 46 NR%, but for the next 300 MY from the amniote ancestor to the present they evolved at the average rate of only 15 NR%. The fastest rate was about 109 NR%. It occurred in the early vertebrates between 500 and 400 MY ago and took place during the evolutionary period when gene duplications first separated the vertebrate myoglobin and hemoglobin branches and then subdivided the hemoglobin branch into alpha and beta hemoglobin lineages. After the α-β duplication which preceded the teleost-tetrapod separation, the alpha lineage initially evolved more rapidly than the beta lineage, but between the tetrapod and amniote ancestors the rate of beta evolution increased several fold. Then in the descent from the amniote ancestor to chicken and mammals an abrupt slowing of both alpha and beta evolution occurred. In the early mammals rates again accelerated but later became extremely slow in some lineages such as the human.

These variations in rates in which we see the pattern of initial acceleration and eventual deceleration can be explained on selectionist grounds, but first we need to consider briefly some major structural and functional differences among globins. The insect, annelid, and mollusc globins are monomeric proteins. Each has a single heme group and a high affinity for oxygen, as described by a hyperbolic shape to the oxygen equilibrium curve. Vertebrate myoglobins are also monomeric globins with high oxygen affinity. The agnatha lamprey hemoglobin, however, has less constant affinity for oxygen and is only monomeric in the oxygenated state (Hendrickson and Love, 1971). Lamprey hemoglobin molecules, unlike most monomeric hemoglobins, exhibit a Bohr effect;

FIGURE 6. Parsimony phylogenetic tree of 55 contemporary globins with well-determined amino acid sequences. The figure is adapted from Goodman *et al.* (1975). The numbers of the links (the nucleotide replacement values) are italicized where the augmentation algorithm has been used to correct for superimposed mutations.

i.e. they release oxygen from their heme iron atoms more readily at reduced pH. Moreover they form homodimers and transitory homotetramers in the deoxygenated state with acidity favoring aggregation. Teleost and tetrapod hemoglobins are heterotetramers, each containing two identical alpha chains, or α subunits, and two identical beta chains, or β subunits. Heterotetrameric hemoglobin shows allosteric or cooperative behavior, also called heme-heme interaction. This behavior is such that the release of oxygen from the first one or two or the four heme iron atoms triggers a conformational change in the spatial relations of the four subunits which greatly speeds unloading the remaining oxygen. Thus, compared with monomeric hemoglobin the heterotetrameric type with its sigmoid oxygen equilibrium curve and Bohr effect unloads oxygen more efficiently in respiring tissues.

Presumably a monomeric hemoglobin with the hyperbolic equilibrium curve, i.e., with strong affinity for oxygen, adequately served our pre-Cambrian ancestors who, as usually pictured, were creeping animals of wormlike size. The direction of evolution which ensued in the vertebrates, however, was toward large animals with high mobility. This direction of evolution required a different type of hemoglobin, one with a sigmoid oxygen equilibrium curve and Bohr effect. It would be expected, therefore, that positive natural selection, rather than the stabilizing kind, would have acted on hemoglobin genes in the early vertebrates to bring about the new form of hemoglobin.

However, why should positive selection speed up evolutionary rates of proteins and stabilizing selection slow them down? Why should the long-run trend, as seen in the descent of hemoglobin genes in the vertebrates, be toward slower rates? The reason is that the very success of natural selection in improving the fitness of competing organisms must, sooner or later, decrease the proportion of mutations which are advantageous and increase the proportion which are detrimental in each surviving and competing lineage. When mutations first produce a potentially useful protein, the possibilities for improvements are considerable. Thus mutations in the gene coding for the protein will often be advantageous and selected for; but after the protein has been perfected, a much larger proportion of mutations would be detrimental and selected against. Once many proteins have been optimized, i.e., more intricately coadapted for the cooperative execution of their functions, molecular evolution would slow down. Such a pattern, however, could be temporarily interrupted by periods of accelerated evolution, such as occurred in the early vertebrates. Mutations were selected in duplicated genes, which multiplied the number of proteins with different functions. Rates of molecular evolution accelerated while selection was optimizing the structures of these new proteins, but eventually the stabilizing component of selection slowed the further rate of change.

Lampreys belong to the most primitive class of vertebrates and have a more complicated hemoglobin than a simple monomer; thus the earliest vertebrates probably also had a more complicated hemoglobin. The emergence of a homotetramer with rudiments of cooperativity could have been a precondition for the selection of functionally superior heterotetrameric hemoglobin. It would explain why the amino acids involved in the interchain contacts of the heterotetramer occur in alpha and beta chains mostly at homologous residue positions. Another fact suggests that the postulated ancestral homotetramer had subunits more beta-like in polymer-forming properties than alpha-like, namely that beta chains often form homotetramers but alpha chains do not (Benesch and Benesch, 1974). The existence of an ancestral beta-like homotetramer would also explain the variations which occurred initially in evolutionary rates between alpha and beta lineages. It appears that following β-α hemoglobin gene duplication, positive natural selection acted on the nascent α locus while stabilizing selection was stronger at the β locus. However once the older β_4 type homotetramer was replaced by the heterotetramer with a specialized alpha chain, positive selection for a more differentiated beta chain could intensify. The rate of beta evolution accelerated between the tetrapod and amniote ancestors, and functionally superior tetrameric hemoglobin emerged. Then α and β loci were both subjected to intense stabilizing selection. The role of natural selection in hemoglobin evolution is clearly revealed by the data summarized in Tables I to III for the amounts of mutational change in different functional groups of residue position in the reconstructed globin sequences during various evolutionary periods.

Table I shows the nucleotide replacement values from the invertebrate-vertebrate ancestor through the basal vertebrate and myoglobin-hemoglobin and β-α ancestors to the basal amniote alpha and beta sequences. The most slowly evolving positions are those which now function in both alpha and beta hemoglobin chains as heme contacts and which must have been well established as such as far back as 600 to 700 MY ago. Yet these heme contacts, although conservative compared with other residue positions, show a 7 times faster evolutionary rate in the preamniotes than in the amniotes. (Compare Tables I to III.) This suggests that heme-heme cooperative interactions were not fully perfected until about 300 MY ago. Positive natural selection in the emerging vertebrates is indicated by the high proportion of changes at positions which now function as $\alpha_1\beta_2$ (and $\alpha_2\beta_1$) contacts in both alpha and beta chains. These contacts favor heme-heme interactions by placing the hemes of the contacting chains in close proximity. Positions which acquired this function evolved 4 times faster than the average rate over all positions. Between the myoglobin-hemoglobin and β-α ancestors, the

TABLE I. Nucleotide replacement lengths for groups of residue positions during:

	Emergence of Hypothetical Homopolymer						*Evolution of Nascent Heterotetramer*			
	Invert.-Vert. to Vert. Anc.		Vert. to Myo-Hb Anc.		Myo-Hb to β-α Anc.		β-α to Amniote α Anc.		β-α to Amniote β Anc.	
Type of residue positions	No. Pos.	Av. Length	No. Pos.	Av. Length	No. Pos.	Av. Length	No. Pos.	Av. Length	No. Pos.	Av. Length
Heme contacts in both β and α and in myoglobin chains	18	0.11	18	0.17	18	0.06	18	0.17	18	0.11
Nonheme $\alpha_1\beta_2$ contacts in both β and α chains	9	0.78	9	0.00	9	0.22	9	0.67	9	0.44
Like-chain contacts in both β and α chains	3	0.00	3	0.00	3	0.00	3	0.33	3	0.67
$\alpha_1\beta_1$ contacts in both β and α chains	13	0.23	13	0.23	13	0.38	13	0.31	13	0.31
With defined functions in α but not β chains	7	0.14	7	0.14	7	0.29	7	0.86	7	0.14
With defined functions in β but not α chains	12	0.33	12	0.08	12	0.17	9	0.11	12	0.67
Remaining interior positions	19	0.16	19	0.26	19	0.05	18	0.22	19	0.26
Remaining exterior positions	67	0.13	67	0.21	67	0.31	64.5	0.48	65	0.28
All positions	148	0.20	148	0.18	148	0.23	141.5	0.40	146	0.30

Positions are grouped according to their functional roles in present-day hemoglobin chains; unaugmented nucleotide replacements for the positions in each group during each period of descent are averaged.

TABLE II. Effect of functions on nucleotide replacement lengths in the amniote ancestor-to-contemporary α lineages

Type of residue positions	No. positions	Av. length	Nucleotide replacements per position per 100 MY
Heme contacts	19	0.37	0.02
Nonheme contact $\alpha_1\beta_2$ contacts	10	0.20	0.01
Salt bridges, associated Bohr effect, $\alpha\alpha$ contact	7	0.14	0.01
Nonsalt bridge $\alpha_1\beta_1$ contacts	14	2.00	0.13
Remaining interior positions	19	1.47	0.09
Remaining exterior positions	72	2.38	0.15
All positions	141	1.68	0.11

Positions are grouped according to their functional roles; unaugmented nucleotide replacements over the amniote (chicken-mammal) α region of the globin genealogy are averaged for the positions in each group. The times of the connecting links over the amniote region when added up represent about 1.6 billion years.

sites which now function as $\alpha_1\beta_1$ (and $\alpha_2\beta_2$) contacts in both alpha and beta chains evolved at twice the average rate of all positions. Many of the amino acid changes at these prospective $\alpha_1\beta_2$ and $\alpha_1\beta_1$ interchain contact sites as deduced from the parsimony reconstruction either persisted or mutated to similar amino acids in later amniote hemoglobin

TABLE III. Effect of functions on nucleotide replacement lengths in the amniote ancestor-to-contemporary β lineages

Type of residue positions	No. positions	Av. length	Nucleotide replacements per position per 100 MY
Heme contacts	21	0.43	0.02
Nonheme contact $\alpha_1\beta_2$ contacts	10	0.40	0.02
Salt bridges, associated Bohr effect, $\beta\beta$ contact	4	0.00	0.00
2,3-DPG binding	4	1.75	0.10
Nonsalt bridge $\alpha_1\beta_1$ contacts	16	2.81	0.16
Remaining interior positions	21	1.57	0.09
Remaining exterior positions	70	3.34	0.20
All positions	146	2.27	0.13

Positions are grouped according to their functional roles; unaugmented nucleotide replacements over the amniote (chicken-mammal) β region of the globin genealogy are averaged for the positions in each group. The times of the connecting links over the amniote β region when added up represent about 1.8 billion years.

chains. This finding supports the notion that the ancestral β-α gene was coding for homotetrameric hemoglobin even before it duplicated.

The positions which are $\alpha_1\beta_2$ contacts in the two hemoglobin chain types evolved at a faster than average rate in the nascent alpha and beta lineages (Table I). Positions which acquired important functions in alpha chains but not in beta evolved in the alpha lineage at a faster rate than other positions. Similarly, positions which acquired important functions in beta chains but not in alpha evolved in the beta lineage at a faster rate than other positions. Clearly, mutations were fixed in the diverging alpha and beta genes because they produced functionally superior hemoglobin molecules. By the time amniote tetrapods evolved, heterotetramers with modern allosteric properties had replaced the older models. Thereafter natural selection acted against further amino acid substitutions at the residue positions involved in heme and $\alpha_1\beta_2$ contacts and the Bohr effect. Nucleotide replacements occurred in the codons for these positions at only one-fifth to one-tenth the average rate (Tables II and III). Moreover the average rate itself was much reduced.

In the early therian mammals, according to the parsimony reconstruction, hemoglobin evolution accelerated again, largely because of amino acid changes at sites at the surface of the hemoglobin molecule which had no sharply defined functions. Perhaps these surface changes occurred because they escaped the scrutiny of natural selection, but we need not wholly accept such a neutralist interpretation. Having fixed the features in the hemoglobin tetramer responsible for its crucial functions, natural selection would then be able to shape the finer adaptations of the protein. These adaptations should especially involve the surface sites because of the sites' potential for interacting with other macromolecules. Amino acid changes in these surface positions can be viewed as further steps in an optimization process in which the functions of proteins become more effectively coordinated. I contend that such optimization happened in the descent of the higher primates and that because it happened, stabilizing selection, more intense than in previous periods, inevitably ensued. This would account for the present decelerated rates of sequence evolution in hominoids.

SUGGESTED READINGS

Many books dealing with primate evolution have been published. The following are recommended, starting with the most recent one: W. P. Luckett and F. S. Szalay (eds.) *Phylogeny of the Primates: A Multidisciplinary Approach,* Plenum Press, New York, 1975. B. G. Campbell, *Human Evolution,* second ed., Aldine, Chicago, 1974; F. S. Hulse, *The Human Species,* second ed., Random House New York, 1971; W. Howells, *Mankind in the Making,* revised ed., Doubleday, Garden City, New York, 1967; J. Buettner-Janusch, *Origins of Man,* Wiley,

New York, 1966; S. L. Washburn (ed.), *Classification and Human Evolution,* Wenner-Gren Foundation, New York, 1963.

A classical book, now largely dated, on the evolution of proteins and other macromolecules is T. H. Jukes, *Molecules and Evolution,* Columbia University Press, New York, 1966.

An extensive compilation of amino acid sequences of proteins can be found in M. O. Dayhoff, *Atlas of Protein Sequence and Structure,* vol. 5 and suppl. 1, National Biomedical Research Foundation, Silver Spring, Maryland, 1972 and 1973.

CHAPTER 10

MOLECULAR EVOLUTIONARY CLOCKS

WALTER M. FITCH

Is there a molecular evolutionary clock? The forces of selection could well drive nucleotide substitutions at such a variable rate that there would be nothing that could usefully be defined as a clock. Nevertheless, many observations have suggested that point mutations occur and are fixed in the genome at reasonably regular rates. To answer the question raised in any statistically valid way requires one to examine the requirements of a clock and its calibration and compare our basic biological data to those requirements. Our conclusion will be that we have several sloppy clocks whose relative merits will be evaluated.

THE PARTS OF THE CLOCK

The first requirement for a clock is a measurable event that occurs with regularity. There are two simple types of regularity, metronomic and stochastic. The former provides the greater accuracy, but no one expects evolutionary phenomena to be metronomic. Indeed, until recently no one had even dared, in the face of selection, to suggest publicly that stochastic regularity might exist. The requirement has three components: events, measurability, and regularity. Each will be examined in turn.

Events

Only two types of events have been given major consideration as components of a molecular clock. The earliest was amino acid replacement in the descent of proteins of various taxa; the latest is nucleotide substitution in the descent of genes of various taxa. The solving of the genetic code permitted an intermediate type in which amino acid replacements were used to infer nucleotide substitutions. (It is common practice in genetics to speak of a gene substitution when a newly arising allele spreads through a population and replaces the previous wild type. In keeping with that practice and as an aid to distinguishing genotypic from phenotypic events, I shall speak of nucleotide *substitutions* and amino acid *replacements*. Both terms imply a population-wide change and should be clearly distinguished from the initial event in an individual, i.e., a *mutation*.) All such information depends upon comparing data from two or more taxa. The nature of the observational data varies.

Measurability

Many proteins have had their amino acid sequences determined; and whenever there is a set of orthologous sequences, amino acid replacement data may be obtained. (*Homology* is here defined as similarity arising by virtue of common ancestry, as opposed to analogy which is similarity arising by virtue of convergent evolution. There are two kinds of homology that are important to distinguish. Two different gene products existing in the same organism may have a common ancestral gene that, either through gene duplication or translocation, gave rise to independent genes evolving side by side in parallel in a single line of descent. Such genes are said to be *paralogous*. Two different gene products whose difference is a consequence of independence arising from speciation are said to be *orthologous* because there is an exact phyletic correspondence between the history of the genes and the history of the taxa from which they derive. Many statements in the literature about homologous genes are untrue except when restricted to one or the other category of homology.) This kind of data is necessary for the inferring of nucleotide substitutions from amino acid replacements. A second but indirect source of information arises from immunological studies where it has been shown (Prager *et al.,* 1972) that the extent (I) of reduced immunological cross reactivity is a monotonically increasing function (linear with respect to log I) of the number of amino acid differences. There is generally little point in trying to infer the number of amino acid differences, and 100 log I is generally used directly as a measure of the number of evolutionary events separating a protein from the orthologous antigen to which the antibody was raised.

Few genes are known in terms of their nucleotide sequences, and therefore substitutions cannot yet be examined directly. The major exceptions are transfer RNAs and ribosomal RNAs. In those cases possessing an orthologous set of sequences, there is little or no paleontological record with which to cross-check the accuracy of any conclusions about regularity. The tRNAs are mostly paralogous. An indirect method of determining nucleotide substitutions is DNA hybridization, where it has been shown that if two DNA strands are not perfectly complementary, they melt or dissociate at a lower temperature than the native DNA and that the degree of lowering is directly proportional to the extent of nucleotide differences.

An important issue is the accuracy with which these differences are measured. (This is apart from the concept of validity which I believe is not an issue because the properties measured do possess a functional relationship to the number of substitutions or replacements wanted.) This issue may be subdivided into reliability, sensitivity, and reciprocity.

Reliability. By reliability I mean the extent to which one can rely upon the fundamental measurement. This subsumes the concept of reproducibility. Given two amino acid sequences, the number of differences may be stated exactly, but the reliability is not therefore 100 percent because the sequences themselves may be in error. It is impossible to give a good value to the extent of this error because it is greatly dependent upon two factors, one of which is the care with which the sequence is determined; this is subject to considerable variation in spite of recent, marvelous technological advances. The second factor is the extent to which the investigators are willing to assume that two peptides of the same composition have an identical sequence. There are many valid reasons for making this assumption but, as the choice among different molecular phylogenies often rests on a small number of crucial differences, the use of partially sequenced sequences is, in my judgment, fraught with more peril than many realize. This is particularly true where the composition includes more than one dicarboxylic acid, not all of whose amidation states are the same, because composition cannot clarify the assignment of the amides. An excellent example of this is the sheep adrenocorticotropic hormone sequence, which was originally asserted, without sequencing, to be identical to the pig sequence in positions 25 and 33 but which had to be subsequently corrected because aspartate-25 and the glutamine-33 of the pig were in fact asparagine and glutamate, respectively, in the sheep (Li, 1972). Amino acids 159 and 162 of subtilisin are threonine and serine in the BPN′ strain and are serine and threonine, respectively, in the Carlsberg strain; amino acids 182 and 183 are serine and asparagine in BPN′ and the reverse in Carlsberg, and the same pair in the same order are also interchanged in positions 248 and 252 of these two sequences (Smith *et al.,* 1966). The last four amino acids of cytochrome *c* are identical in rattlesnake, pigeon, and duck, but they show different orders (Nolan and Margoliash, 1968).

It came as something of a shock to me that delta hemoglobin has never been sequenced; only a few of its peptides have been. Thus not only does the alignment of one delta chain to another depend upon the assumption that unchanged composition between two orthologous peptides implies an unchanged sequence, but the entire sequence depends upon that same assumption holding between the paralogous peptides of human beta and delta hemoglobin which, having arisen by gene duplication, possibly as long as 45 MY ago, have therefore had perhaps as much as 90 MY of separate evolution.

For those wishing a crude estimate of the trust they may invest in amino acid sequence data, I would suggest that the error rate for a *completely* sequenced protein is perhaps 2 percent with a value only half that if it is reasonably close to another completely sequenced orthologous protein. Where the sequence is guessed at by homologizing compositions of peptides, the error rate is probably at least doubled, although it clearly depends upon the protein's rate of divergence, the time since divergence of the solid sequence used as the basis for homologizing, and the size of the peptide.

The other two measures, immunological distance and hybridization, are general measures and therefore do not suffer any error in determining the sequences, but the very fact that they are general measures means that they have the disadvantage of being an overall average rather than a precise knowledge of the amount of change. On the other hand, a process averaged over a genome may be considerably better than a precise value from a single locus that could be unrepresentative of the genome.

It is important, however, to know whether immunological distance is linear with respect to amino acid replacements because it is here a fundamental assumption that this is so. Figure 1 presents the data show-

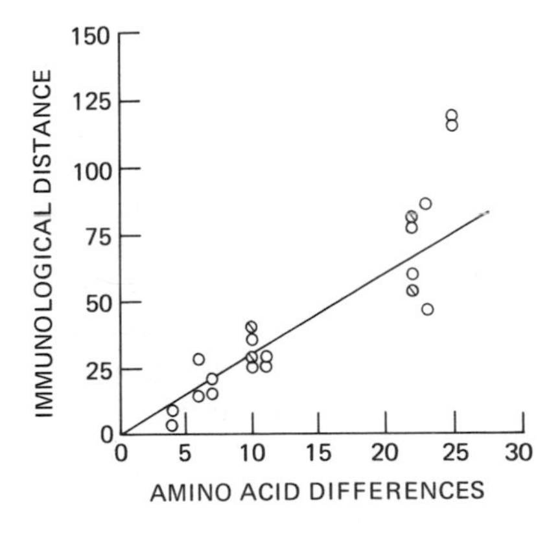

FIGURE 1. Relation between amino acid differences and immunological distance for lysozyme. Data for amino acid differences come from Prager *et al.* (1972), while data for immunological distance come from Prager and Wilson (1971). Data for a given amino acid difference come in multiples of 2 because the test is performed in both directions. Where there are two pairs the second one is designated with a slash across the circle.

ing that there is a linear relationship between the two, at least up to an immunological distance of 70. The slope will differ from one protein to another, but that is not a problem so far as regularity of rate is concerned. The variance associated with this slope suggests a source of uncertainty regarding statements about rates.

The same issue arises for hybridization where the slope has been assumed to be 1.5 percent base-pair mismatches per 1°C lowering of the melting point. Again, the slope itself is not crucial to the issue of regularity, but, as with the immunological distances, the scatter of the points away from the line suggests a considerable uncertainty as to the percent of mismatched base pairs even if the melting-point lowering were known without error. Kohne (1970) gives this latter error as ±0.4°C.

Sensitivity. By sensitivity I mean the extent to which one can pick up minor variants, rather than its other meaning, the ratio of the absolute error to the magnitude of the variable being measured. In this respect all three of the methods—sequences, immunological distances, and hybridization—are insensitive. Variants amounting to less than 10 percent of population are unlikely to be recognized in pooled samples by sequencing techniques. The other two techniques are even less sensitive. Although this insensitivity is unfortunate in trying to answer some questions, such as the extent of heterozygosity, it is nearly irrelevant to the examination of evolutionary rates in the present context.

Reciprocity. By reciprocity I mean the extent to which the evolutionary distance from taxon *A* to taxon *B* is the same as from *B* to *A*. Reciprocity is perfect in the case of sequence comparisons because the number of amino acid differences (or the inferred nucleotide differences) is not a function of the direction of measurement between the taxa.

Reciprocity is a significant element in immunological tests because the degree to which an antibody formed against protein *A* reacts with protein *B* may not be the same as the degree to which an antibody formed against *B* reacts with *A*. The best data of which I am aware come from Wilson's laboratory, where the standard deviation of the reciprocal pair distances from their mean, expressed as a percent of that mean, is 15 percent for bufonoid albumins, 20 percent for bird lysozymes (Prager and Wilson, 1971), and 14 percent for bacterial azurins. The full extent of the lack of reciprocity is partially masked in these figures because one utilizes numbers obtained by taking the logarithms of the observational data, i.e., of the index of dissimilarity. One extreme, but nevertheless unnerving, example is the human-baboon lysozyme distance which is 66 in one direction and 127 in the other (Wilson and Prager, 1974). This means that the index of dissimilarity differed by a factor of more than 4. The logarithmic transformation is also necessary to obtain the linearity shown in Figure 1.

The degree of reciprocity for hybridization is unclear. The melting point of DNA composed of one strand from each of two different species might be expected to be the same regardless of which species' DNA contained the radioactive label and which species' DNA was used in excess. Nevertheless, it is clear that when done both ways the two melting points of the heterologous DNAs differed by 1.5°C when the two species were man and green monkey (*Cercopithecus aethiops*) (Kohne, 1970) and by 1.0°C when they were a junco (*Junco hyemalis hyemalis*) and a thrush (*Catharus guttatus*) (Shields and Straus, 1975). The difference was only 0.2°C when satellite DNA from *Mus musculus musculus* and *M. caroli* were paired in both directions (Rice and Straus, 1973). It is unclear whether the smaller error in the last instance is due to the use of the highly repetitive satellite DNA or to the test being applied intraspecifically. It should be noted that because the man-monkey melting-point lowering is 10.5°C, a potential error of 1.5°C is nearly 15 percent of the observation.

Regularity

The requirement for a clock is a measurable event that occurs with regularity. We have so far discussed the events, amino acid and nucleotide changes, and their measurement by sequencing, immunology, and hybridization. Because the events only occur on the time scale of millennia, their regularity can only be examined under some evolutionary hypothesis that distributes the events to the various historical intervals. This involves using the set of pairwise distances between the taxa in the manner shown in Figure 2. The phylogeny is here assumed to be as shown, and the problem is to determine how to apportion the distances among the three legs of the tree that join the taxa. Because the three desired quantities summed in three different ways yield the three known pairwise distances, their value is easily obtained as shown on the figure. Note that the fact that legs a and b are unequal despite equal times for divergence clearly demonstrates that equality of rates is not a built-in assumption in assigning leg lengths by the procedure of Fitch and Margoliash (1967).

If there are more than three taxa, the process is repeated as many times as there are bifurcating nodes, the only difference being that A, B, and C become collections of one or more taxa and the three distances become the average of all the pairwise distances between the members of the sets. Given t taxa, there are $t - 2$ such bifurcating nodes, and the results are easily combined to assign values to each segment on the tree. If, for example, in the data in Figure 2 A is a set of two taxa

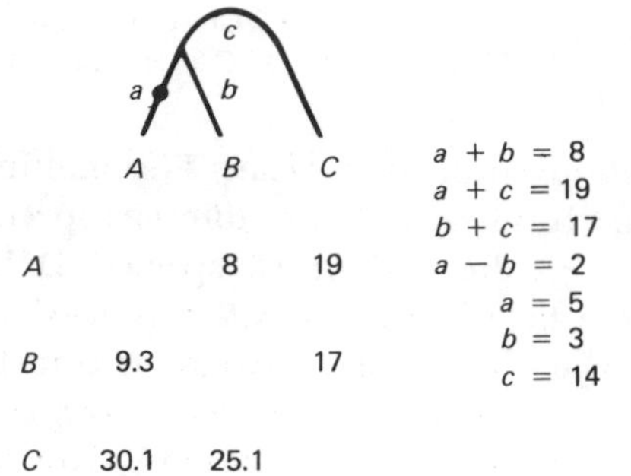

FIGURE 2. Assigning values for the amount of change to the various legs of a phylogenetic tree. It is assumed that one has some measure of distance (dissimilarity) between all possible pairs of taxa being examined. The upper right portion of the matrix gives a hypothetical example for three taxa, *A, B,* and *C.* The tree above the matrix shows legs of length *a, b,* and *c* connecting the taxa to a common node. The problem is to determine the values of *a, b,* and *c* if one is given the matrix values. The relationship between the given and desired data is shown by example in the equations on the right together with their solution. The procedure applies equally well if *A, B,* and *C* are sets of taxa and the matrix values are average distances. The lower left portion of the matrix shows the number of amino acid replacements one would expect to have occurred in order to observe the data in the upper right half if the later were the number of amino acid differences observed between 30 variable positions. For the lower-left data, the values of *a, b,* and *c* are 7.15, 2.15, and 22.85, respectively.

which, having been assigned the roles of A' and B' in a previous calculation, then gave a' and b' values of 3 and 1, the present leg, $a = 5$, would have its own bifurcation at the dot shown on that leg in the figure. The dot is located at a height of 2 because that is the average of 1 and 3, which means that the part of the leg above the dot has a length of 3 $(5 - 2)$. Repeating this process successively with nodes more and more removed from the present time will quickly give the needed data for the entire tree.

While it is sufficient for the present purposes to accept the phylogeny from sources outside the data, it should be clear that the data themselves could be the source of the tree. For example, the joining of *A* and *B* as in Figure 2 could have been chosen on the basis that *A* and *B* are more similar to each other than either of them is to *C*. (Other numerical taxonomic procedures for constructing phylogenies from such data may be found in Sneath and Sokal, 1973.) Whether to let the data themselves determine the phylogeny is not irrelevant to the regularity of the clock. Clearly, any other joining in Figure 2 than the one shown, say *B* to *C*, cannot affect the values of *a, b,* and *c;* but now it would be legs *b* and *c* that would have the same evolutionary time span, and the difference in their rates, 3 and 14, would be less likely (compared with 3 and 5) to be explainable as stochastic variation. This poses something of a dilemma because using the data to determine the *structure* of the tree by the method of Fitch and Margoliash (1967) or its derivatives tends to make the rates somewhat more uniform than they would otherwise be. On the other hand, if the data were a better basis for inferring the true phylogeny, then using any other basis for testing the null hypothesis, that the evolutionary rates are uniform, is to bias the test in favor of rejection of the hypothesis. If, however, one rejects that hypothesis even after using the data to determine the phylogeny in the above fashion, that conclusion is of the strongest possible form because any bias introduced works against rejection.

VARIOUS TREATMENTS OF AMINO ACID SEQUENCE DATA

It is now useful to examine the various ways of treating amino acid sequence data in the light of the preceding method of assigning change to the legs of a tree. In Figure 3 we see an amino acid for some hypothetical position of a protein sequence from each of four taxa. In practice one examines many positions, and only rarely would a single position provide a case this extreme, but the simplicity obtained makes for an easy comprehension of several important comparisons.

Amino acid differences

The upper portion of Figure 3 shows the distance matrix in the form of amino acid differences, and the phylogeny shows how the changes would be assigned to the legs by the method just outlined. The assignment is "perfect" in the sense that the sum of the values on the legs connecting any two taxa is equal to the number of differences shown in the matrix. This is not generally true where more than three taxa

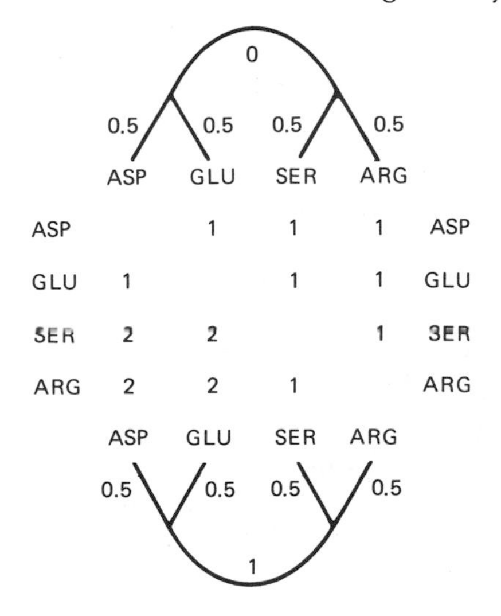

FIGURE 3. Assigned leg values for a simple phylogeny. The data are a single amino acid for four taxa and an assumed phylogenetic relation. The upper right-hand portion of the matrix gives the number the amino acid differences between all pairs, and the numbers of the upper tree show the values assigned to the legs using the procedure in Figure 2. The lower left-hand portion of the matrix is for the same data, but the distances are the minimal number of base differences between the codons of the amino acids, and the lower tree shows the corresponding leg values assigned.

are examined because the number of observations in the distance matrix will then exceed the number of leg values to be estimated from them; i.e., the problem is overdetermined. Because all pairwise distances are the same, all possible phylogenies would permit equally good fits to the observed differences.

Minimum coding distances

In the lower portion of Figure 3 the distance matrix is based upon minimal nucleotide differences in the codons. The conversion of a codon for aspartate or glutamate into one for arginine or serine will in every conceivable case require at least two nucleotide changes. Thus these distances are now 2 rather than 1, as in the previous case. The values assigned to the legs of the phylogeny are also changed so that an additional evolutionary change is shown and the total number has increased from two to three. In addition the phylogeny shown is the only one for which the assignment of values to the legs remains perfect. It is also the only tree consistent with the proposition that the most similar taxa are the most closely related. Moreover, the trees based on these data, requiring a total of three changes rather than two as seen in those based on amino acid differences, are clearly closer to the truth because four different amino acids necessarily require at least three changes of state. For all these reasons, the example, despite its contrived nature, clearly shows that important information can be lost by ignoring the relation between the amino acids and their genetic coding.

It is possible to carry the process two steps further. In order to get the Ser-Glu distance of 2, their respective codons would need to be UCR and GAR; whereas in order to get the Ser-Arg distance of 1, their respective codons would need to be AGY and CGY (or AGR, as discussed later). But notice that the serine codon representation is not the same in both cases, whereas it must in reality possess a single representation. We can improve our procedure by requiring a unique representation to be used in all comparisons. The problem, however, is which representation to choose. An arbitrary but unbiased procedure is to use that form which is most similar to the closest phyletic relative. If the phylogeny is not predetermined, the distance matrix itself must be the source of the choice. We do not have other positions to help with the choice, but it proves unnecessary because the lower part of the matrix in Figure 3 is unbiased with respect to codon choice and clearly Ser is closer to Arg than to Asp or Glu. Thus we must assign AGY to Ser. This also restricts the six arginine codons to CGY and AGR. The distance matrix of the upper portion of Figure 4 is based upon the AGY codon for serine and the CGY codon for arginine, and the assignment of leg values shows a further increase in total number of changes from three to three and one-half. Note how restricting serine to the

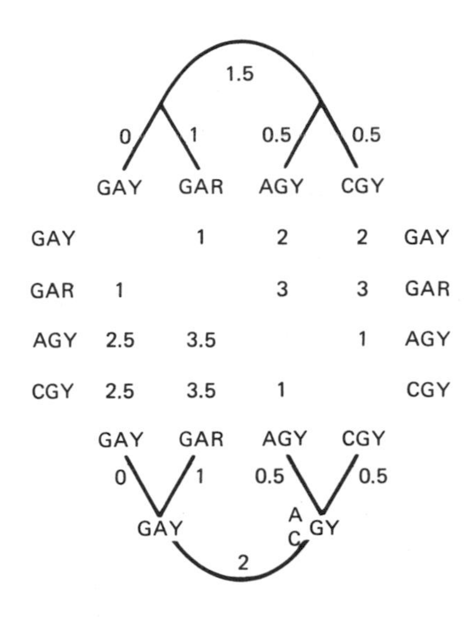

FIGURE 4. Assigned leg values for a simple phylogeny. The basic data are identical to the data in Figure 3, except that specific codons have been assigned to serine and arginine. The reasons therefor are given in the text. The upper right-hand matrix is for the nucleotide differences shown, and the tree above shows the leg values assigned. The lower tree is one of several most parsimonious trees, and the values assigned to the legs by that technique have been summed to provide the data in the lower left-hand matrix. R stands for purine (A or G), Y stands for pyrimidine (C or U).

AGY codon necessarily makes the distance to the glutamate codon, GAR, equal to 3, a considerable change from the single amino acid difference. The assignment of values is still "perfect." Few if any investigators have attempted to refine the data in this fashion, probably because at this point one might as well attempt the phyletic process illustrated in the lower half of Figure 4.

Parsimony

The phyletic process asks what the fewest number of changes of state is that will account for the observed data in the context of a given phylogeny. The procedure for finding this answer was given by Fitch (1971*b*) and proved by Hartigan (1973). The answer for the nucleotide sequences shown, which is not dependent upon a matrix of distances, is given in the matrix and therefore, unlike the other three cases, is a product of, rather than the raw material for, the phylogeny. The assignment of values to legs cannot therefore be judged perfect or imperfect because the standard is now parsimony rather than agreement to a matrix of pairwise differences. The sum of the assigned values is now 4, the largest total yet.

It should be noted that assigning arginine to the AGR codon rather than CGY provides an equally parsimonious result, four changes of state, and the values assigned to the legs will vary slightly. The phylogeny shown, however, remains the best by the parsimony criterion. No other unrooted phylogeny, regardless of form or the choice of codons, will permit the explanation of these four amino acids in as few as four nucleotide substitutions. In any event it should be clear that phyletic distances, although dependent upon the phylogeny assumed, are greater than pairwise distances that do not concern themselves with the ancestral relationships and that compared with any of the procedures based solely upon pairwise differences the parsimony process ends up detecting more change. Because even that is minimal, its greater magnitude is at least one obvious virtue to parsimony procedures. As such, parsimony does no violence to biological thought. To explain change economically is not to assume that nature necessarily proceeded precisely as outlined. Parsimony procedures cannnot be applied to the immunological or hybridization procedures; therefore, they may not be the most sensitive for testing the regularity of evolutionary events.

Correcting for multiple changes at a single site.

The minimal nature of all the preceding values suggests that a proper evaluation of the regularity of the evolutionary events may require a further correction. There are four such corrections presently available. The first is simply a Poisson correction. Because changes occurring in a position already changed once are not observable in a comparison of two sequences, we can ask how many changes would have had to have occurred in order to see the number of differences observed? If 80 percent of the positions are unchanged, then $r = -\ln 0.8 = 0.223$; so $0.223/0.2 = 1.12$ times as many changes would have occurred as positions are different (see Margoliash and Fitch, 1968; and Dickerson, 1971). The correction can only be applied to immunological data if the relation between immunological distance and the fraction of the protein sequence altered is known. It can even be applied to parsimony data by correcting the values assigned to each leg of a tree, because each such number is necessarily the number of differences between the sequences at the nodes on either end of the leg.

The second procedure, by Dayhoff (1972), attempts to allow for the obvious bias in the nature of the alternative amino acids allowed by evolution to occupy a given position. Although otherwise similar in principle as a problem to the previous, its solution was obtained by a Monte Carlo procedure and suggests that the Poisson correction does not add enough changes for one to estimate properly the total number of evolutionary changes.

The third procedure (Holmquist, 1972) attempts to determine all nucleotide changes, not just those that change the encoded amino acid. It also suggests that the Poisson correction undercorrects.

The fourth procedure (Goodman *et al.*, 1974*b*; Chapter 9) augments the nucleotide substitutions in each interval of the most parsimonious tree so that long unbranched intervals will have numbers of substitutions comparable to those found in many-branched intervals involving the same total evolutionary time.

With two exceptions (Goodman *et al.*, 1974*b*; Jukes and Holmquist, 1972) the above corrections have not to my knowledge been used prior to a test of the regularity of evolutionary rates. It is, moreover, unclear that any of these corrections would increase the power of the test because it is conceivable that any resulting increase in the difference between the evolutionary changes on the two descending sides of a node is more than compensated by the larger values of the difference between and by a larger uncertainty in the estimate of those values. Moreover, the corrections should only be applied to those positions where change is evolutionarily acceptable (the covarions) and there is considerable evidence that their number can be considerably less than the entire gene and, worse, that those positions appear to change in the course of evolution (Fitch, 1971*a*). For these reasons, the corrections will not be further considered here, where the question is the regularity of the events. Corrections must of course be employed if the rate itself is being estimated.

CALIBRATION-DEPENDENT TESTS OF THE CLOCK HYPOTHESIS

We have been considering the first requirement of regular, measurable events. The second requirement is an accurate set of calibration times (although we will see later a way to avoid this requirement). Consider Figure 5 to represent for some data set a phylogenetic solution where the bifurcating nodes are placed at heights representing the average number of substitutions/replacements separating the two sets of taxa that descend on either side of those nodes. A least-squares fit of those heights against the known times of divergences should provide an average rate of change per unit time, and that rate multiplied by the

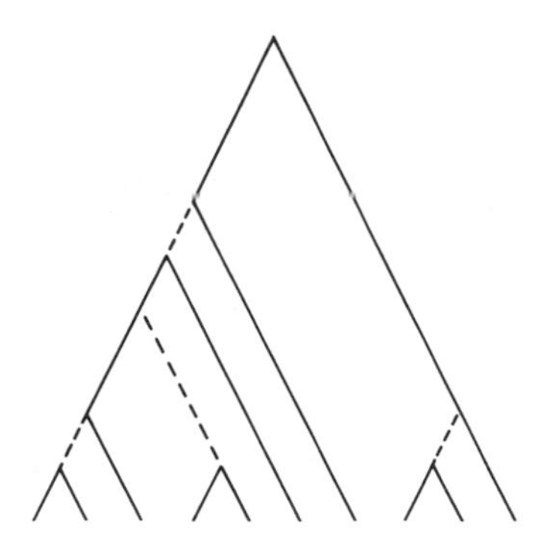

FIGURE 5. Independent evolutionary pairs. The tree when disconnected at the dotted segments leaves the taxa in pairs with no evolutionary history in common between any two pairs. Such pairs are a prerequisite for independent samplings of the evolutionary process.

time of divergence for any node should give the expected amount of change for which the observed height is an estimate. For a Poisson distributed process, such as radioactive decay, the variance (σ^2) of the estimate is equal to the estimate. Thus a test of the clock hypothesis is potentially available.

It should first be pointed out, however, that there is a bias in the data. Much of the evolutionary change accounted for in the descent from any given node is common to the change surrounding an adjacent node. In other words, the estimates are not independent of each other. It is, however, a simple matter to disconnect such a tree into components such that no pair of taxa share a common lineage. This is shown in Figure 5 by dotted lines whose removal properly disconnects the tree. Restricting our nodal estimates to those that represent the common ancestors of the connected pairs of taxa reduces the observations by nearly a half; but they are now independent, and the above noted bias is removed.

All tests involving such data pose a dilemma if we should reject the null hypothesis that substitutions/replacements accumulate randomly over time. That rejection may result from the failure of any of the assumptions; one of those assumptions is that we know precisely the times of divergence of the taxa involved. We could not then be certain whether the clock hypothesis or the set of divergence times was in error. We shall therefore not consider further such calibration-dependent tests except to note several important studies of this type. These include those of Ohta and Kimura (1971) on hemoglobin and cytochrome *c* sequences, Kohne (1970) on DNA hybridization, and Wilson and his colleagues (Wallace *et al.*, 1971; Sarich, 1972*a*; Maxson *et al.*, 1975) on immunological distances.

CALIBRATION-FREE TESTS OF THE CLOCK HYPOTHESIS

We turn now to calibration-free tests of the null hypothesis; the null hypothesis states that amino acid replacements (or nucleotide substitutions) accumulate at a stochastically uniform rate over geological time. In other words, if the data in Figure 2 are either replacements or substitutions, we can ask whether leg lengths a and b (5 and 3) are within the expected range of random variation given only that they are independent estimates of that same rate. Note that we do not need to know the time of divergence, because the common ancestor occurred the same number of years ago for both A and B. This is easily tested by a chi-square test, where $\chi_1^2 = (a - b)^2/(a + b)$, which in this case is 0.5 and clearly not significant; i.e., we cannot reject the hypothesis.

The difficulty is that one never observes the number of substitutions or replacements, only differences. If we wish to know how many substitutions or replacements were required to see as many differences as were observed, we might apply the Poisson correction. It should be noted that this requires the assumption that all positions are equally likely to receive the next change, an assumption known to be incorrect (Fitch, 1971*a*). It further requires the assumption that a position identical in both sequences has always been that way since their common ancestor. It also has the effect of increasing the significance of the difference in the rates in the two legs. If, for example, the data in Figure 2 are the number of amino acid differences between two proteins of length 30 (or longer but possessing only 30 variable positions), the observational difference values in Figure 2 of 8, 19, and 17 become replacement values of 9.3, 30.1, and 25.1, respectively. The difference between a and b was $19 - 17 = 2$, but it now becomes $30.1 - 25.1 = 5$; so $\chi_1^2 = 5^2/9.3 = 2.7$, which is many times larger than that for the original data. It is not clear that the larger χ^2 is not at least partially artificial, because the values 30.1 and 25.1, rather than being observations, are now estimates whose values have a variance not included in our consideration and because the numerator is the difference between two large numbers that have always been increased disproportionately (which means that the difference between the number of events on the two legs usually increases more than the total length of both legs!). As a consequence, although the total numbers increase, the shorter leg usually decreases in length. This is a disturbing result. A proper statistical treatment of this problem has not been done. If the original data were immunological distances, there is yet another complication, in that they should properly be converted to amino acid differences for which another variance is introduced but up to now has not been included.

The use of such derived-rate data for any statistical test of the null hypothesis, that rates are uniform over time, must include within that hypothesis the assumption of equal variability of all the codons so that even if a statistically significant departure from expectation occurred, it could be attributed to the failure of the equal-variability assumption rather than to any real inequality of rates.

The procedures outlined for a pair of taxa (but using a third, *C*, taxon to get the data) can be applied to a tree for a larger number of taxa provided that the independence of the data is maintained by choosing pairs that contain no common paths of descent. This would be accomplished as discussed earlier in regard to Figure 5 except that one would have to exclude the pair involving the ultimate ancestor because the number of changes on either side of that ancestor are not estimated in the absence of an assumption of equal rates in both lines. In view of the problems just enumerated, neither this nor any of the other procedures seems to have yet provided a sound statistical test of the clock.

In cases where statistics are hard to do properly, it is sometimes better to choose an example that is convincing without statistics. This was done by Jukes and Holmquist (1972) using cytochrome *c* and comparing the taxa snake, bird, and turtle. If we assign them to *A, B,* and *C,* respectively, of Figure 2 (with replacement distances of 58, 55, and 11), we obtain leg values for *a, b,* and *c* of 51, 7, and 4, respectively. Now the difference between 51 and 7 is compellingly unequal. However, if we let the data dictate the phylogeny, we interchange *A* and *C,* associating the turtle rather than the snake with the birds. This is, of course, a considerable distortion of our current biological viewpoint, but the result is that the length of legs *a* and *b* is now 4 and 7, respectively, and *c* becomes 51, and a significance in rates flies out the window because $(7 - 4)^2/(7 + 4) = 0.82$ is an acceptable χ^2 for one degree of freedom. I, for one, am sufficiently impressed with the general tendency toward equal rates and with the extreme nature of this result to believe that the truth really is that either one or more of the sequences is incorrect, that our current view of amniote phylogeny is incorrect, or both. I incline toward the third explanation. The mystery ought to be sufficiently important to entice the sequencers to tackle this area with greater industry.

There is one other method (Langley and Fitch, 1973, 1974) of asking the question about rates that depends upon the parsimony procedure. In this case, seven proteins were examined from 17 mammals. The procedure finds the minimum number of nucleotide substitutions to account for the descent of the protein sequences from a common ancestor and assigns them to various legs of the phylogeny as required by the parsimony constraint. One then attempts to find by a maximum-likelihood procedure the evolutionary rate for each protein such that the expected number of substitutions is as close to the observed as possible. The likelihood of the observed outcome, as well as that of the best possible observations, is calculable, and the natural logarithm of their ratio is a chi-square value with one degree of freedom. Moreover, it is possible to test two independent propositions. One is that the overall (combined) rate of change is uniform over time, the other is that the relative (one protein to another) rate is uniform over time. It should be clear that it is possible for the overall rate to be constant but with significant increases in evolutionary rates of some proteins compensated by significant decreases in the rates of others. Alternatively, the overall rate could change significantly without any significant change in the relative rates if all the proteins changed their rates proportionately.

The results of this test are shown in Table I. It can be seen that there is a significant deviation in both the overall and relative rates

TABLE I. Tests of hypotheses

	Corrected χ^2	d.f.	p
Among proteins within legs (relative rates)	166.3	123	6×10^{-2}
Among legs over proteins (total rates)	82.4	31	4×10^{-6}
Total	248.7	154	6×10^{-6}

d.f. is degrees of freedom, and p is the probability that results this far removed from expectation would arise by chance if the null hypothesis were true, i.e., if amino acid changing nucleotide substitutions are accumulating uniformly over time.

of evolutionary change. Moreover, the relative rates are from the same lines of descent and therefore cannot be explained by a generation-time effect. Thus, the use of geological rather than generation time cannot explain the deviation from uniformity of rates. Furthermore, the test does not require knowledge of paleontological dates; so errors there cannot account for the result. Finally, because the estimates were corrected for nonlinearity due to hidden multiple substitutions at a single locus, the result is not due to this potential source of error either. We must conclude that the clock, at least for amino acid changing nucleotide substitutions, is not the stochastic timepiece that radioactive decay is.

How bad is the clock? The chi-square value is about twice the degrees of freedom; hence the variance of our clock is about twice that expected. For radioactivity measurements, the estimated standard deviation, s, is the square root of counts; the variance is equal to the counts. For 100 counts, $s = 10$, or 10 percent; so for 200 nucleotide substitutions $s = \sqrt{400} = 20$, which is also 10 percent. Said another way, a doubling of the variance implies that to achieve the same accuracy the sample size must be twice what one would need for radioactivity measurements. For example, a time based on only eight such nucleotide substitutions would have a standard deviation equal to 50 percent of the estimate. Nevertheless, averaged over larger time intervals and greater numbers of proteins the estimates ought to be fairly good and, if so, to be linear with respect to paleontological time. That this is in fact the case is shown in Figure 6. Thus the clock is significantly erratic, but averaged over enough events the variability tends to be washed out. This statement is probably true for other clocks than the one being tested here, but the confidence one can have in the others has not been properly ascertained.

It might be observed that a measure linear in differences rather than linear in changes ought not in principle to be linear with time as

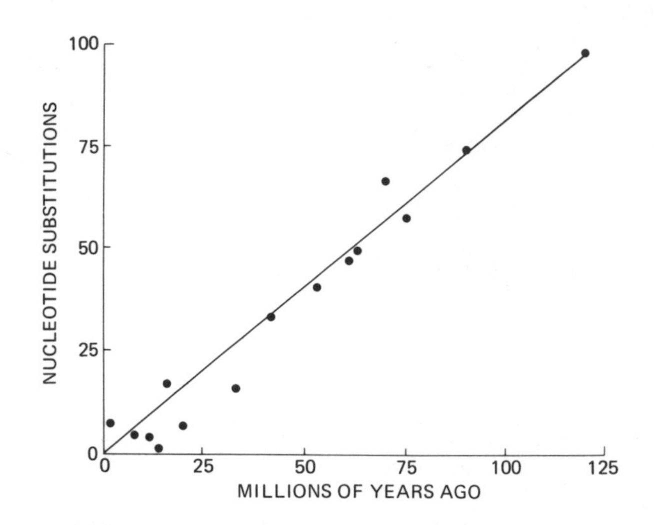

FIGURE 6. Nucleotide substitutions versus paleontological time. (L. Van Valen and M. Goodman provided the dates of divergence for a phylogeny of species of 17 mammals.) Nucleotide substitutions are the number expected on the basis of a maximum-likelihood fit to the maximum-parsimony solution for seven gene products from these species. viz., cytochrome *c*, fibrinopeptides A and B, α and β hemoglobins, insulin C-peptide, and myoglobin. (The line was simply drawn through the outermost point, which represents the marsupial-placental divergence.)

well. An amino acid change has a probability $p = 1.0$ of being a difference when the two proteins are identical before the change but only a $p = 0.8$ of being a difference if they are already 20 percent different. The p value is even lower if some positions cannot survive a mutation there. Thus, for example, if immunological distances or percent nucleotide differences are indeed linear with geological time (apart from small distances), then either evolutionary rates are not only not uniform over time and must have been slowing down in a manner precisely compensating for what would otherwise require some form of the Poisson correction, or there is an error somewhere. Of course, if there were the requisite compensating change in evolutionary rates, such a measure does become a suitable device for a clock. However I would be worried at my luck in finding so fortuitous a compensation.

EVOLUTIONARY RATES

As long as Figure 6 suggests that we have a clock, it is useful to ask what the rate of evolution is. In the present case, the slope of the line is 98.17 amino acid changing nucleotide substitutions in all seven proteins per 2×120 MY of evolution. Because these seven sequences consist of 578 amino acid positions, or 1734 nucleotide positions, the evolutionary rate is 0.5×10^{-9} amino acid changing nucleotide substitutions per year per nucleotide position. For fibrinopeptide B, the rate is 1.7×10^{-9} amino acid changing nucleotide substitutions per year per nucleotide position. Kohne (1970) on the basis of his nonrepetitive DNA hybridization data between the anthropoids and the old world monkeys (Cercopithecidae) found 4.9 percent base differences and, using a divergence time between 30 and 45 MY ago, one can calculate an average genomic substitution rate at between 1.0 and 1.5×10^{-9} substitutions per year per nucleotide position. This is not greatly above that from the overall amino acid changing rate; the difference would be even less if the base differences per degree Celsius were 1 rather than 1.5 percent. This suggests reasonably strong selective constraints on nonrepetitive DNA outside the region of the structural gene, especially considering the fibrinopeptide B rate. This is even more true in view of the silent substitution rate shown in Table II.

Three proteins, α and β hemoglobin and histone $F2a_1$, have had portions of their messenger RNA sequenced from two different taxa.

TABLE II. Rates of silent nucleotide substitution

	α	β	$F2a_1$	Total
A: Codons (third position)	11	19	18	48
B: Number of *A* unchanged	5	12	12	29
C: Substitutions per site $[-\ln (B/A)]$	0.788	0.460	0.405	0.504
D: MY of evolution	122	122	120	121
E: Rate_1 $[C/(D \times 10^{-3})]$	6.45	3.77	3.38	4.17
F: Rate_2 $(E/0.716)$	9.02	5.27	4.72	5.82

A gives the number of codons whose third-position nucleotide is known in both species. *B* gives the number of third-position nucleotides that are identical. *C* is the average number of changes per site (B/A) corrected for sites containing multiple changes by using the Poisson distribution, where $B/A = e^{-r}$ and r is the corrected value wanted. *D* is twice the number of years since the common ancestor of the two species. *E* is the rate in silent nucleotide substitutions per third-position nucleotide per 10^3 MY. *F* is a lower bound of the mutation rate in mutations per nucleotide per 10^3 MY if we assume that all third-codon-position silent mutations are neutral and 0.716 of all third-position mutations are silent. From Forget *et al.* (1974, 1975) and from unpublished data kindly provided by the following persons: B. G. Forget for human hemoglobin mRNA; W. Salser for rabbit hemoglobin mRNA; M. Grunstein for sea urchin histone mRNA.

We may ask about differences in the third nucleotide position that do not affect the amino acids encoded, and which I am here calling *silent* substitutions. As the table shows, with due allowance for multiple substitutions per site, the rate of silent substitutions is 4.2×10^{-9} silent nucleotide substitutions per third nucleotide position per year, or nearly one order of magnitude greater than those that change the encoded amino acid. We can also estimate a lower bound on the mutation rate. We know from Kimura (1968) that the substitution rate for neutral variants equals the neutral mutation rate. If all the silent substitutions are neutral, then the preceding rate is the neutral mutation rate. If there is selection operating, then the mutation rate is in fact greater than the substitution rate; hence the estimate is a lower bound. However, only 0.716 of all third-position variants are silent; so the mutation rate must be 1.4 times the silent substitution rate. That this is a rather low estimate is suggested by the fact that if this were the true mutation rate, then approximately 10 percent of all amino acid changing mutations would have to be evolutionarily acceptable to account for a substitution rate only one order of magnitude less. But if we must increase the mutation rate in order to lower the fraction of acceptable mutations, then we must admit that selection is operating to constrain the choice among silent third-position alternatives.

SUGGESTED READINGS

The constancy of evolutionary rates follows from the neutrality theory of molecular evolution as proposed by M. Kimura, Evolutionary rate at the molecular level, *Nature,* **217,** 624–626, 1968; and by J. L. King and T. H. Jukes, Non-Darwinian evolution: random fixation for selectively neutral alleles, *Science,* **164,** 788–798, 1969. Among the many critiques of the theory, a brief but pointed one is R. C. Richmond, Non-Darwinian evolution: a critique, *Nature,* **225,** 1025–1028, 1970. The reader interested further in the problem of molecular evolutionary rates should consult T. Uzzell and D. Pilbeam, Phyletic divergence rates of hominoid primates: a comparison of fossil and molecular data, *Evolution,* **25,** 615–635, 1971.

Clear and short discussions of methods to construct phylogenetic trees are W. M. Fitch and E. Margoliash, The construction of phylogenetic trees, *Science,* **155,** 279–284, 1967; M. O. Dayhoff, Computer analysis of protein evolution, *Sci. Amer.,* **221**(1), 86–95, 1969; and G. W. Moore, J. Barnabas, and M. Goodman, A method for constructing maximum parsimony ancestral amino acid sequences on a given network, *J. Theor. Biol.,* **38,** 459–485, 1973.

CHAPTER 11

EVOLUTION OF GENOME SIZE

RALPH HINEGARDNER

Increases and decreases in the amount of cellular DNA have played a major role in the evolution of organisms. In fact, because of the nature of the genetic mechanism these changes along with changes in nucleotide sequence are the only two DNA modifications that are important contributors to the evolutionary process. Other alterations, such as the appearance of a new and different nucleotide, a codon with other than three nucleotides, or a different chemical basis for the genetic system, are all essentially excluded. Any of these would have to be accompanied by such major alterations in the entire makeup of the organism that it is very unlikely that if they would occur, an organism would be able to survive through the transition period while still successfully competing with organisms of the standard design. In other words, the transition organisms would have a selective disadvantage. Individual nucleotides can be modified once in position, but these always start as one of the fundamental four. The study of evolution at the DNA level is the study of DNA gain and loss and of nucleotide changes.

CHANGES IN NUCLEOTIDE SEQUENCE

There is no room to doubt that nucleotide sequences have changed over time. This can be seen as differences in the arrangement of amino acids in similar proteins of different species (Chapters 9 and 10) or in the nucleotide sequences underlying proteins. In many of these studies the degree of variation nicely corroborates evolutionary relationships derived from anatomical and paleontological evidence. In time, studies of the amino acid sequences, or DNA and RNA nucleotide sequences, should be able to resolve what are now ambiguous evolutionary relationships as well as produce new insights into the entire process of evolution.

Before we continue further, it is best to introduce and clarify the following terms:

Generalized and *specialized* will refer for the most part to anatomical features. Generalized will be used to describe organisms that share numerous features with other members of their taxon. In contrast, a specialized organism shares fewer features with the members of its taxon and will differ from them in the presence or absence of certain features. For example, most parasites are specialized. They are often missing many parts typical of their free-living relatives, and they often have unique adaptations to their form of life. In fish, the sea horse is specialized; the trout or salmon is generalized. The bat is a specialized mammal; the rat is generalized.

Lineage. It is difficult to conceive of a species when evolution over long time spans is being considered. A species is a set of similar or interbreeding organisms that occupy a thin slice of time. The slice we are most familiar with is the present. The organisms that do exist in the present have ancestors who in turn had ancestors back ultimately to the origin of life. Together these form a lineage, and all present or fossil organisms are parts of lineages. Lineage will be used here to designate these time-related organisms.

EVOLUTION

A group of organisms first appears in the fossil record as a small number of related species, often of rather generalized form, or at least generalized relative to what some of the descendants will look like. If the group prospers, there is diversification into an ever-increasing number of different forms. This is the period of rapid evolutionary radiation. Phyletic branching continues for some time but eventually with reduced speed. There are now numerous adaptations to the general environment the group lives in; the group has entered a plateau period. During this period the number of new species that arise approximates the number that becomes extinct. Around this time extreme specialization becomes more and more typical of new species; generalized forms are less abundant. Gradually the group enters a post-plateau period where specialization is typical; the total number of species diminishes, and the group becomes progressively less dominant. Finally, the entire group may become extinct or remain as a small number of highly specialized species.

There are innumerable examples of this process and its consequences. The trilobites, jawless fishes, ammonites, brachiopods, dinosaurs, many of the ferns, and mammals are just a few illustrations of the pattern followed by evolution above the species level. Some of these groups have completely disappeared, others nearly so, and some remain. The mammals are now probably living in a post-plateau period. Bernard Rensch (1959) was one of the first to describe this process in detail. It is interesting that much of this process can be simulated rather easily on a computer (Raup *et al.,* 1973).

DNA CONTENT AND EVOLUTION

Differences in amount of DNA between species have received relatively less attention than amino acid sequence studies, though a large amount of data has accumulated over the past quarter century. Reviews of DNA content and evolution can be found in Ohno (1970), Smith (1974), and Hinegardner and Rosen (1972). In this chapter, eukaryotic DNA content will be given as the haploid amount.

The first broad survey of DNA content in animals was carried out by Mirsky and Ris in 1951. They only examined about 60 species scattered throughout the animal kingdom, but they were able to cautiously suggest three relationships between DNA content and evolution: (1) There might be an increase in DNA content going from the lower invertebrates, such as the sponges and coelentrates, to the higher invertebrates. (2) Related organisms, such as members of the same family, tend to have similar amounts of DNA. (3) Evolution of the land vertebrates may have been accompanied by DNA decreases. These relationships have held up with time, and the pattern they observed for the land vertebrates has been found in other groups as well.

DNA content is now known from direct measurements for close to a thousand species of animals and plants and from indirect measurements, such as calculations from chromosome or nuclear size, for many additional species. The direct measurements have been made with the techniques of colorometric or fluorometric determination of the DNA in a known number of cells, and microspectrophotometry of single Feulgen stained cells. Both approaches give results that are fairly close though seldom identical. As a rough estimate, almost all assays, including the indirect measurements, can be considered to be within 30 percent of the actual DNA value. The more careful direct analyses should be within 5 to 15 percent of the actual value.

DNA amounts in free-living organisms (i.e., not viruses) span a range from 0.007 pg [1 picogram (pg) = 10^{-12} g] for an average bacterium to 100 pg or more for the haploid amount of DNA in some plants and salamanders (Sparrow *et al.,* 1972). The high-DNA-content organisms have more than 10,000 times the DNA of the bacteria. Most eukaryotic animals have approximately 1 to 5 pg of DNA per haploid cell, or about 300 times more than a bacterium. Even among the eukaryotes the range is large. The fungi are at the lower end with an average 0.09 pg. Because the upper end is 100 pg or so, this gives a range covering three orders of magnitude. Table I gives a summary of DNA amounts in some major groups of organisms. There are a number of characteristics and correlations involving DNA content, including the following:

1. DNA distribution within a major group such as a phylum or class is usually asymmetrical, with more organisms at the lower end of the distribution. Figure 1 illustrates this for the teleost fishes, one of the most extensively assayed groups. The figure represents assays for almost 300 different species. The same general shape can be seen in the distribution of DNA content in the annelids

TABLE I. Range of (haploid) DNA content per cell in various groups of organisms

	DNA, haploid (pg)	Reference
Bacteria		Wallace and Morowitz, 1973
Mycoplasmata	0.0017–0.0037	
Other	0.0033–0.01	
Animals		
Sponges	≈ .05	Mirsky and Ris, 1951
Coelentrates	0.35–0.73	Hinegardner, unpublished
Annelids	0.09–5.3	Conner *et al.*, 1972
Crustaceans	0.09–15.8	Bachmann and Rheinsmith, 1973 Rheinsmith *et al.*, 1974
Insects	0.1–7.5	Sparrow *et al.*, 1972
Molluscs		Hinegardner, 1974*a*
Gastropods	0.43–4.4	
Bivalves	0.65–5.4	
Echinoderms	0.54–3.3	Hinegardner, 1974*b*
Lower chordates	0.20–1.5	Hinegardner, unpublished
Sharks and rays	2.8–9.8	Hinegardner, unpublished
Teleost fish	0.39–4.4	Hinegardner and Rosen, 1972
Amphibians		Bachmann *et al.*, 1972 Olmo, 1973 Hinegardner, unpublished
Urodels	19 to ≈ 100	
Anurans	1.2–7.9	
Reptiles	1.5–3.5	Hinegardner, unpublished
Birds	1.7–2.3	Bachmann *et al.*, 1972
Mammals	3.0–5.8	Bachmann, 1972
Plants		Sparrow *et al.*, 1972
Psilopsids	129–313	
Ferns	6.0 to ≈ 100	
Gymnosperms	4.2–50	
Angiosperms	1.0–89	

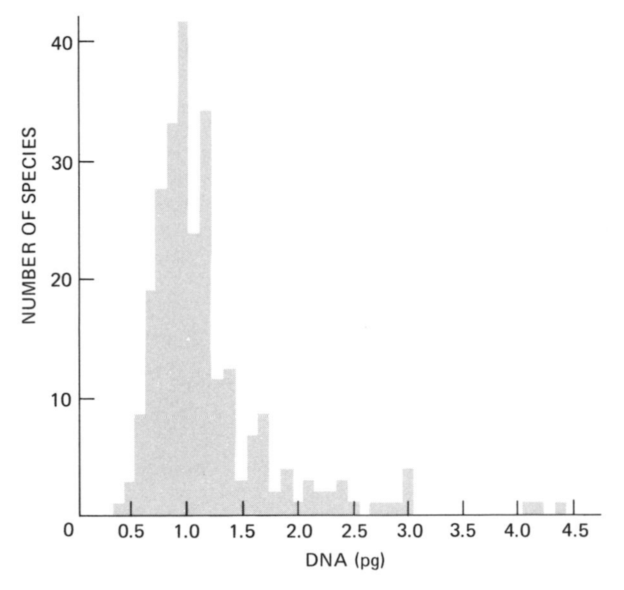

FIGURE 1. Distribution of the haploid DNA content in teleost fishes. The skewed distribution is common to many groups of organisms.

(Conner *et al.,* 1972), in at least two classes of molluscs (Hinegardner, 1974*a*), the echinoderms (Hinegardner, 1974*b*), crustaceans (Bachmann and Rheinsmith, 1973; Rheinsmith *et al.,* 1974), amphibians (Bachmann *et al.,* 1972), and mammals (Bachmann, 1972). In many instances the distribution approximates a logarithmic normal curve (Bachmann *et al.,* 1972). Though this shape may not be universal, it is very common.

2. DNA content may or may not correlate with chromosome number. In plants it is fairly common to find that the two do correlate, sometimes in consistent ploidy series. In animals the correlation is less frequent, and there are many examples of wide differences in DNA content accompanied by little or no change in chromosome number. The echinoderms have DNA in amounts ranging from 0.5 to over 3 pg (Hinegardner, 1974*b*), yet all chromosome numbers so far determined for the phylum are either 21 or 22 (Colombera, 1974). Assayed salamanders of the genus *Plethodon* all have 14 chromosomes, but DNA ranges from 18 to 69 pg (Mizuno and Macgregor, 1974). In mammals there is no correlation between DNA content and chromosome number (Bachmann, 1972). There are also instances of correlation. The fishes and molluscs with larger amounts of DNA generally have more chromosomes (Hinegardner and Rosen, 1972; Hinegardner, 1974*a*).

3. The higher eukaryotes seldom have less than 0.4 pg of DNA (Figure 2). This suggests that below a certain minimum number DNA is not easily lost, or at least that organisms at the very low end of the group do not survive

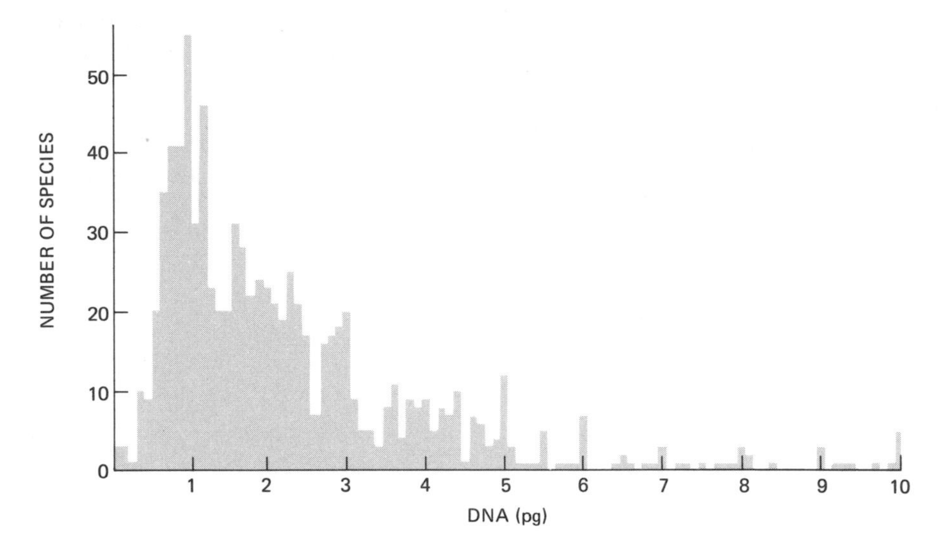

FIGURE 2. The distribution of DNA content among all assayed eukaryotes with DNA amounts between 0.1 and 10 pg.

long. A comparison of the spread (coefficient of variation) in DNA content among members of fish families versus amount of DNA supports this belief (Hinegardner and Rosen, 1972). There is a significantly smaller percentage difference in DNA content among members of families that have low amounts of DNA.

4. Adult body size and DNA content are positively correlated in the molluscs (Hinegardner, 1974*a*). *Drosophila virilis* is at least twice as large as *D. melanogaster* and has twice as much DNA (Endow and Gall, 1975). Polyploid plants have more DNA and often are larger in size than their diploid progenitors.

5. DNA content correlates with specialization in many different groups of organisms. Low-DNA members of a group tend to be more specialized (Stebbins, 1966; Hinegardner, 1968; Bachmann *et al.,* 1972). Figure 3 illustrates this. It is easy to see that the fishes on the left side are for the most part "fishey" looking fishes; they have the shape someone would draw if he were asked to draw a fish. These tend to be generalized fishes. Most animals on the right are specialized; they lack the classic fish shape. This same correlation is apparent in the insects (Bier and Müller, 1969), amphibians (Bachmann *et al.,* 1972), molluscs (Hinegardner, 1974*a*), sea cucumbers (Hinegardner, 1974*b*), mammals (Bachmann, 1972), and several plant groups (Stebbins, 1966). Specialization and low DNA content are common features of organisms.

6. Animals and plants that are considered to represent primitive or ancient and relatively slowly evolving lineages often have more or much more DNA than the average of their particular taxon. Some examples are lung fish, with 80 to 140 pg (Pedersen, 1971), the coelacanth, 7 pg (Cimino and Bahr, 1974), salamanders, 20 to 100 pg (Sparrow *et al.,* 1972), and the psilopsids and ferns, about 20 to over 200 pg (Sparrow *et al.,* 1972). Organisms which are ancient but highly specialized, such as the lampreys and hagfish, have fairly low amounts

of DNA. These two average about 2 pg, which is still high if one considers their degree of specialization.

7. The average DNA content of a taxonomic group such as a class or order appears to be inversely correlated with the number of species in that group. This is particularly apparent in the vertebrates, where the speciose teleost fishes and birds have the least amount of DNA (Bachmann, *et al.,* 1972).

8. DNA content and nuclear and cell size are positively correlated; the more DNA, the bigger the cell and nucleus (Commoner, 1964; Baetcke *et al.,* 1967). The correlation is good enough to allow one to make rough estimates of DNA content simply by measuring cell or nuclear size.

It is unlikely that knowledge of a single parameter, such as DNA content, will allow one to say much about the details of evolution, such as which species are most closely related or how fast a group has evolved. Rather, DNA measurements relate more to the major features of the evolutionary process, where time spans in millions of years are involved and where major groupings of organisms are being considered.

The first primeval living thing undoubtedly had less DNA per cell than even the simplest nonparasitic prokaryote living today and certainly less than the higher eukaryotes. Over the history of life on Earth DNA content per cell has increased, though not in a continuous fashion as we shall see. To produce the up-and-down distribution seen in Table I, there must have been times when DNA content increased and other times when it decreased. Both of these movements have played major roles in the process of evolution.

DNA DECREASES

Some groups of organisms appear to be unable to accommodate DNA increases, or at least there seem to be no contemporary members of the group that have significantly more DNA than the average. A few examples are the mammals, neoteleost fishes (the modern fishes), reptiles, sea urchins and sea stars, limpets, and jellyfish. Over time and in the absence of DNA increases, two things will happen to the DNA in such organisms—the nucleotide sequences will change, and the amount of DNA will decrease. The chance that the amount of DNA will remain the same is probably about equal to the chance that the sequence will remain unchanged, which is practically zero. However, the rate of loss and the rate of change may not be the same.

The consequence of DNA change and loss is evolution toward specialization and eventual extinction. The only changes available to an organism will be modifications of already existing components and therefore existing nucleotide sequences. A particular species seldom exists for long periods of time. The physical environment fluctuates, and lineages that survive follow these fluctuations by changing the amount or kinds of components. Ultimately, this comes down to changes in their DNA.

Van Valen (1973) gives an important second reason for changes—the Red Queen hypothesis. The Red Queen is the woman who is running to stay in place in Lewis Carroll's novel *Through the Looking Glass.* Evolution is an analogous thing. There is always selection for the organism that is best able to find food and avoid being preyed upon; but because one organism is another's food, selection acts on both,

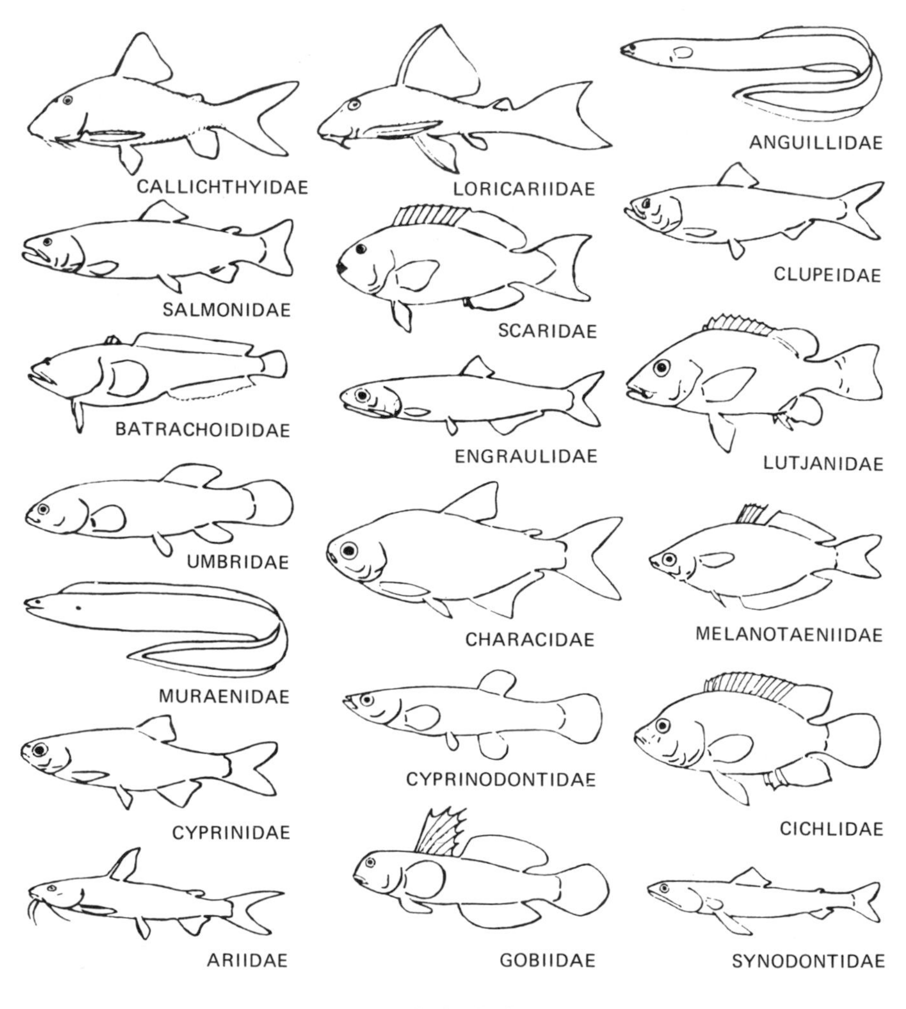

FIGURE 3. A qualitative demonstration of the correlation between low DNA content and specialization. Fish families on the left have high amounts of

predator and prey. For example, each time the survival of the prey improves by better avoiding the predator, there will be an increase in potential prey biomass. With this potential food source around, it is only a matter of time before the original predator type evolves or a new predator arises that is able to eat some of this new biomass. The process then begins all over again. Even though organisms are continually evolving, the average lineage is getting nowhere. Figuratively,

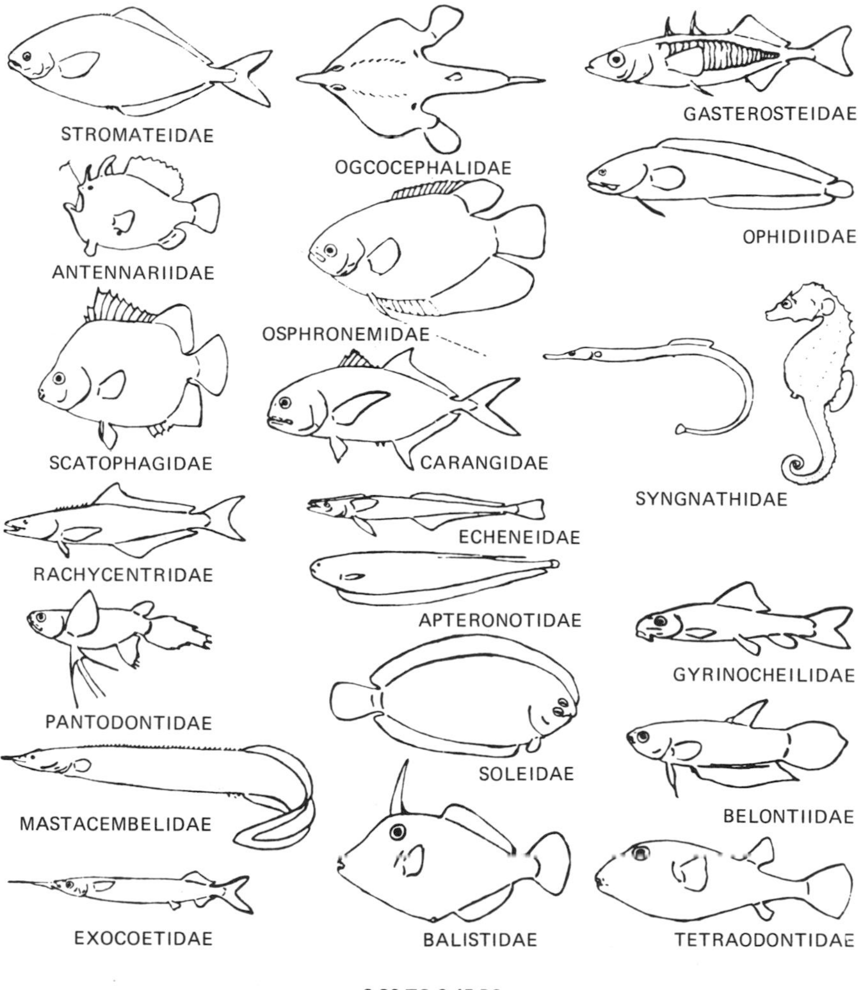

DNA; those on the right have low amounts. DNA content decreases from top to bottom and from left to right.

the lineage is running to stay in place. This is why there are so few "living fossils." The few species that exist are the minority that has not evolved and yet has not fallen so far behind as to become extinct. Survival goes to lineages, not to species.

EVOLUTIONARY EXPEDIENCY AND DNA LOSS

Consider a group of organisms that do not or cannot increase their DNA content. For the purposes of simplifying the discussion, assume that practically all nucleotide sequences play a needed role. Among these organisms, species x_1 will evolve to x_2, a new species. If x_1 has been able to survive so far, it is necessarily a well-integrated being. That is, all parts act together in the activities of the whole. The fact that there are few genes that cannot be made into recessive lethals confirms this. The most expedient, and therefore adaptively probable, change from x_1 to x_2 would be for x_2 to do what x_1 does best, but to do it even better. This carries with it the consequence that x_2 will have to do something else less well or not at all. Nothing, neither machine, intellectual concept, nor organism can indefinitely become better adapted to a specific role and at the same time become better for all roles.

For example a lightly armored fish might evolve into a heavily armored fish, thus becoming a less desirable meal. This is a form of specialization. Along with being less acceptable food, the fish is also heavier and would tend to spend more of its time near the bottom. The portion of the nervous system and other parts of the anatomy that are involved in flotation and fine swimming coordination may gradually be lost. There is no longer selection for these, and in fact there may be selection against them. There is also no selection for the presence of the portion of the DNA which is involved in the formation and control of these structures, and in time that DNA may be lost. The outcome would be an organism with less DNA than its ancestor. Further evolution might lead to the formation of a mouth suitable for bottom feeding. With that, teeth become less important and may be lost. The outcome would be still less DNA. The final organism might be a heavily armored, ornately spined, rather bizarre looking bottom-feeding fish, with a low amount of DNA. At some point this fish lineage might become so specialized that it would be unable to further evolve and even a small environmental shift might be fatal.

Another reason why evolution in the absence of DNA increase leads to specialization is that a set of genes for certain parts of the organism cannot become a set of genes for a very different set of parts. Bones change to bones of another shape, but they are still bones; digestive enzymes have changed amino acid sequences, but they are still digestive enzymes and unlikely to become hair proteins or the like.

As a lineage continues to run, it may succeed in remaining in place, but it also may lose a little DNA with each step. In time it may run out of specializable parts. With this, the lineage is no longer able to stay in place and in time becomes extinct. There are, of course, other routes to extinction, such as the instance where the appropriate specialization never arises, where environmental changes are too rapid, or where an overwhelming predator or parasite appears. These can occur at any time; with lower amounts of DNA a lineage may become increasingly more vulnerable to them.

There are organisms supporting this picture. Deep-sea teleost fishes average significantly more DNA than their shallow-water relatives (Ebeling *et al.,* 1971). The deep sea is a more stable environment than shallow water. The difference in DNA amounts may in part be due to the need for more changes in a shallow-water lineage. Another piece of evidence is that there are few if any very low-DNA organisms that would fall under the general heading of living fossils; living fossils tend to have more than average amounts of DNA. On the other hand, there are many successful recent organisms with low amounts of DNA.

Low DNA values, specialization, and speciation all tend to go together. Specialization carries with it loss of parts; the fewer parts, the less organism there is to accommodate subsequent specialization. Also, each step down the specialization road excludes competition and leads to selection for further specialization. If adaptive radiation is a part of this, then increasing numbers of niches become available for potential species. All this will be accompanied by loss of DNA. As we have seen, at some point expediency is its own terminator.

Drosophila, the mammals, birds, and neoteleost fishes are all animals that are popular subjects for evolutionary studies. All show little evidence of DNA increases, are speciose, and have low to moderate amounts of DNA. These animals appear to be expedient evolvers.

DNA INCREASES

There have been times when new and different types of organisms arose by other than specialization. These have arisen following DNA increases. Were it not that DNA content can increase, life on Earth would have ceased to exist long ago, probably billions of years ago. Those organismal lines of descent that do exist have done so because at times DNA content increased. This means that increases in DNA content may be favored by selection, though such increases need not be everyday occurrences. The evidence indicates that increases happen infrequently and irregularly, with gaps measured in millions of years.

There is a wide range of data demonstrating that DNA has increased.

At the single-gene level there is the amino acid sequence analysis showing that proteins such as the digestive enzymes trypsin, chymotrypsin, and elastase are the products of gene duplication (Shotton and Hartley, 1970). The hemoglobins and myoglobin (Zuckerkandl, 1965) and the NAD-utilizing enzymes (Rossmann *et al.,* 1974) have also arisen by gene duplication. There is evidence as well of internal duplications in single proteins such as ferrodoxin (Eck and Dayhoff, 1966) and the carp calcium-binding protein (McLachlan, 1972). The ubiquitous repetitive DNA that is found throughout the eukaryotes provides another example of DNA duplications (Chapter 12). All these tell of past duplications followed by nucleotide change. There is no reason to believe that similar events are not going on in contemporary organisms. There is also the phenomenon of polyploidization, which is present in many organisms but particularly common in plants. Finally, measurements of haploid DNA amounts in related organisms indicate that members of many groups have been able to prosper with significantly more DNA than their relatives. The evidence for gene duplication has been well documented by Ohno (1970).

Some explanations for the accurrence of DNA increases are: larger cell size, slower cell division and development, slower metabolism, duplication of critical nucleotide sequences, more repetitive DNA, and ultimately improved control. Whether or not all these explanations are correct, DNA increases have occurred so often that increase in itself must have some immediate selective advantage. Whatever the advantages, and there may be several simultaneously, they must be of a general nature. There are innumerable examples of closely related organisms with DNA amounts varying over a wide range, sometimes by 100 percent or more. Conversely, there are groups of organisms whose members all have close to the same amount of DNA, as though increases were excluded, possibly for genetic reasons.

Most organisms fall into four partially overlapping DNA classes, as illustrated in Figure 4:

Class 1, the bacteria. Among the free-living organisms, the bacteria have the lowest amount of DNA. Most have between 0.003 and 0.01 pg. The bacteria of the genus *Mycoplasma* have 0.002 to 0.003 pg (Wallace and Morowitz, 1973).

Class 2, the fungi. DNA amounts in the limited number of species that have been assayed are greater than in the bacteria but less than in almost all higher eukaryotes (Sparrow *et al.,* 1972).

Class 3. Included here are practically all animals and some plants. Within this group there seem to be two subclasses:

a. These organisms are undoubtedly products of DNA duplications but themselves do not seem to be able to further increase their DNA content. These are organisms such as the mammals, birds, reptiles, neoteleost fishes, sea urchins and sea stars, limpets, and jellyfish. Assayed mem-

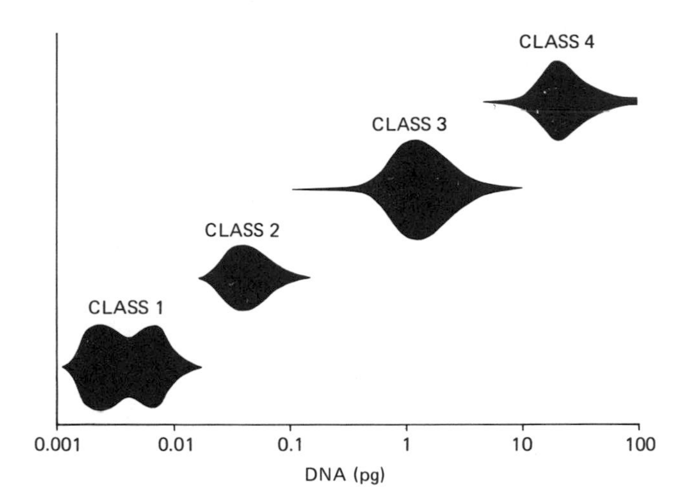

FIGURE 4. Four classes of organisms graphed according to their DNA content. Class 1 includes the bacteria; class 2, the fungi; class 3, almost all animals and some plants; class 4, many plants, salamanders, and some non-teleost fish.

bers of these groups seldom have DNA in high amounts or in amounts a great deal larger than the mean for the group. Further, related organisms usually have close to the same amounts of DNA. Most organisms that fall into this subclass seem to be near the plateau or post-plateau period of their evolution; many are actively speciating. DNA amounts are 0.1 to about 3 pg.

b. Unlike subclass *3a,* members of this subclass show evidence of an ability to increase their DNA content. Related organisms may sometimes differ by 100 percent or so in their DNA content. Most members of this subclass have less than 6 pg of DNA, and none have vast amounts. Most of the molluscan classes, the ostariophysin fishes, serpent stars, sea cucumbers, some polychaete annelids, crustaceans, and some plants belong here.

Class 4. Class 4 is composed of those groups whose members have greater than 10 pg of DNA, sometimes much greater. For example, the salamanders have from 10 to almost 100 pg of DNA. Many plants fall in the same range.

When DNA amounts are separated in this way, it is somewhat easier to see a pattern. The prokaryotic genetic system might preclude large amounts of DNA. Fungi possibly occupy habitats not requiring a complex genetic system. They are also relatively simple organisms. Members of class 3 are more complex organisms. Those in subclass *3a* may be trapped there because their genetic makeup does not tolerate significant increases in DNA. If that is the case, then recent and future evolution should be by expediency. In the case of the sea urchins and sea stars, there has not been much recent evolution. Viewed in terms of the Red Queen hypothesis, the mammals, birds, and the neoteleost fishes are running; the sea urchins and sea stars have found a fairly stable resting place.

Subclass *3b* contains some of the most interesting organisms. Many appear to be in the midst of major evolutionary changes or to be the products of recent evolutionary events. As in subclass *3a*, some belong to lineages that are undergoing extensive specialization. Two examples of what seem to be recent DNA increases will be given. The catfish of the genus *Corydorus* appear to be at the start of significant evolutionary change following a recent gene duplication (Hinegardner and Rosen, 1972). One member of this genus is *C. aeneus*, the common armored aquarium catfish. Almost all assayed members of the genus have more DNA than any other catfish. Furthermore, the amounts vary widely from one species to another. This is also a very speciose group, with approximately 100 different species living in the Amazon basin. They are all anatomically similar to each other, and it is sometimes hard to find distinct boundaries between species. The group may be undergoing evolutionary radiation. Here we also find the fairly common occurrence in animals of the chromosome number being uncorrelated with DNA content.

The neogastropod snails are another and larger group that may now be rapidly evolving. This is a fairly recent order of snails and includes the murex shells, welks, and cones. All assayed species have about twice as much DNA as the other snails (3.2 compared to 1.6 pg). They have atypical body structures and seem to be more complex than the other snails. They also have unique specializations, such as the poison apparatus in the cones. They certainly are a successful group.

At the extreme is class 4, with organisms that use DNA as a form of specialization. If a moderate increase in DNA has a selective advantage, there may be circumstances under which another increase on top of the first is even better, and another better yet, and so on. Of course, at some point increased amounts of DNA may become a burden. The upper limit may be around 100 to 200 pg. Unlike most organisms, these high-DNA species might be the consequence of repeated selection for any benefits that bulk DNA conveys. Usually the animals and some plants that fall into class 4 are members of lineages that have long since passed through the plateau phase of their evolution. Examples are the lung fishes, salamanders, psilopsids, and ferns.

What appears to have happened with the catfishes and neogastropods and what must have happened many times in the past, particularly in subclass *3b* animals and plants, was first a DNA increase. This can happen through a number of possible mechanisms, such as polyploidization, tandem duplications, or unequal crossing over. If the entire DNA compliment were duplicated, the new organism would have double of everything: genes, their control, and whatever else DNA does. If this or any other degree of duplication is in any way selectively advantageous, or at the least neutral, the organism becomes a prime candidate for the start of a new evolutionary line. If the majority of sequences are not needed as precise duplicates, only two things can and inevitably will happen: Nucleotides will be lost, and the sequences will change. Practically all evolution that has led to more complex or markedly different organisms must have begun with DNA duplication.

A different avenue that could lead to increases would be the addition of new nucleotides one at a time, the way a typist constructs a sentence, or by some other addition or insertion process. The resulting sequence would be unrelated to existing sequences. The probability of a usable sequence being generated this way is something like that of an untrained monkey typing a meaningful 100-word paragraph. It is assumed here that this is not an important source of new DNA.

EVOLUTION AFTER DNA INCREASE

Following duplication, the loss or change of a portion of the DNA does not involve the same penalty as it would in a nonduplicating organism. As long as there is at least one working copy of a nucleotide sequence, the other sequence will change. This can be concluded from amino acid sequence analysis of constrained proteins, such as cytochromes *c* versus the much less constrained fibrinopeptides (Margoliash *et al.,* 1971).

In the large group of subclass *3b* and class 4 organisms we have lineages which are simultaneously losing DNA through evolutionary changes, gaining DNA by various forms of duplication, changing nucleotides gradually in constrained sequences, and changing them rapidly in nonconstrained sequences. This gives organisms two types of DNA. There are the selectively constrained sequences. These are the ones that affect events in the organisms, and include genes and their control. This will be called *primary* DNA. Then there are the much less constrained sequences produced by duplication; this is the *secondary* DNA. A fuzzy area undoubtedly lies between the two; however, the two types are probably bigger than the overlap and can be examined as two populations of nucleotide sequences.

The primary DNA is almost certainly the smaller of the two types in an average organism. All the higher eukaryotes have far more DNA than is needed for a reasonable maximum number of genes and their control, yet these organisms exist in abundance, as though having a great amount of nucleotide polymer were advantageous. As has already been pointed out, no organism increases its DNA content indefinitely; there is an upper limit that is different for different groups. There must be a broad optimum range for secondary over primary DNA.

A rough estimate for this range can be made simply on the basis of the number of genes an organism can reasonably be expected to have versus the total DNA content. The estimated range for the number of genes is wide, going from about 10,000 for some of the lower eukaryotes to 400,000 as a reasonable high upper limit. An average protein subunit has a molecular weight of about 50,000 (Klotz, 1967); this would be about 400 amino acids or 1,200 nucleotide pairs. An average multicellular eukaryote has about 2 pg of DNA. This would contain 4×10^9 nucleotide pairs. A 10,000-gene organism would be utilizing 12×10^6 nucleotides, and a 400,000-gene organism 480×10^6 nucleotides. If both organisms contained 2 pg of DNA, the low-gene organism would be using 0.3 percent of its DNA for genes, and the highest-gene organism, 12 percent of its DNA. We must keep in mind, however, that we have at best crude estimates of the number of genes and that we do not know everything DNA does besides code for proteins. There are undoubtedly control sites and spacers (Davidson *et al.*, 1975*b*) and possibly things no one has conceived of. Probably all of these will not occupy more than twice the amount of DNA that codes for proteins. At the maximum then, in our average organism primary DNA accounts for 0.6 to 24 percent of the haploid DNA. DNA amounts range from 0.1 to 100 pg in the higher eukaryotes, which is 1.7×10^8 to 2×10^{11} nucleotide pairs. Even with the lowest amount of DNA, there is enough for 70,000 genes, their control, and anything else DNA does.

Specialized organisms have less DNA than generalized ones. By the previous arguments specialized organisms would also be expected to have fewer genes and other required sequences. If we place the low gene estimate in a low-DNA organism (0.1 pg) and the highest DNA estimate in a high-DNA organism (25 pg, the approximate mode for class 4), we can get some notion of primary versus secondary. This gives 10 and 2 percent, respectively, for genes and other things DNA sequences do directly, or an average of about 6 percent of the DNA being primary, 94 percent secondary.

A reasonable picture of the DNA in an average organism is as follows: first there is a small portion, the primary DNA, with sequences that code for proteins and other aspects of the organism's immediate existence. These are sequences required for the organism's survival in its environment. This is around 6 percent of the DNA. Second, there is a large portion, the secondary DNA, that originated by duplications.

Specific sequences in this portion of the DNA may play only a small role in the organism's activities and at best be weakly maintained by selection. Changes or losses would be more rapid than in the primary DNA. Secondary DNA may arise by duplication of existing secondary DNA as well as by duplication of primary DNA.

We have already seen what can happen in an organism with secondary DNA. In the hypothetical organism that never duplicated its DNA, evolution was toward specialization and loss of DNA. These same forces will be acting in the organism that has duplicated its DNA. In addition, there is a potential source of new working sequences coming from descendants in the secondary DNA of previously working sequences. Most nucleotide changes, wherever they occur, will be useless and sometimes harmful. Infrequently, there must be a few useful new ones. There are many examples of proteins that have arisen from duplications of existing proteins. If the components of an organism are as closely integrated as they seem to be, the origin of a new protein, or anything else, may have little selective advantage because initially it probably will not fit into the tightly integrated flow of biochemical events that already exists. There is no reason to believe organisms are preadapted to the chance origin of a new protein. Selection leads to the optimum use of what already exists.

The process of fitting the products of a new nucleotide sequence into an existing organism may not be as simple as it first seems. As Muller (1950) pointed out, immediately after the origin of a potentially advantageous sequence there is likely to be only weak selection for it, or conceivably there could be none at all. However, if the sequence persists for an extended period, selection might be primarily for modification of other portions of the organism. These changes will gradually integrate the new sequence into the organism. In time a different organism may arise.

Let us use an imaginary biological example: If an enzyme arose in a person that permitted the individual to digest cellulose, a vast new food supply would become available. However, cellulose is associated with hard structures; it is most abundant in plants that do not taste very good to people, such as grass and trees. These may contain compounds that can accumulate as toxins over years of consumption. All this means that the person could eat some selected forms of cellulose but cellulose would probably be a minor dietary component. Nevertheless, when food is scarce or expensive, survival might be modestly enhanced. Selection in the lineage inheriting this gene would be toward additional modifications. Teeth and gut would be expected to become better able to withstand hard material, taste preferences would be modified, and enzymes would arise that could detoxify harmful trace compounds. These new features could arise from portions of the secondary DNA as well as through modification of existing genes. Throughout this process cellulose would become a more and more desirable food source, and in time the enzyme cellulase would become tightly integrated into the flow of events in the body.

There are many other variations on this theme that could be presented. It is likely that more than one integration process is going on at the same time in an evolving lineage. Evolution is an overlapping continuum of these processes. The common feature is that now and then a new nucleotide sequence leads a lineage into a new environment in the form of a new organism. DNA duplication is the pump that drives this form of evolution.

HYPOTHETICAL EVOLUTION OF DNA CONTENT

Figure 5A illustrates schematically the relation between primary and secondary DNA. A small number of secondary DNA modifications produce new sequences that are selected for and become part of the primary DNA. A good portion of the nucleotides are lost from the secondary DNA. Fewer nucleotides are lost from the primary DNA. If the whole system is at equilibrium, duplications balance the large secondary DNA and smaller primary DNA losses. (The curved arrow inside the circles symbolizes sequence change. Again, the rate is faster in the secondary DNA.)

If this picture is approximately correct, then it is possible to design a simple computer model that shows what DNA content might look like in a lineage over long periods of time. In the model it is assumed that there is 7 times as much secondary as primary DNA, that 0.043 of the secondary DNA and 0.009 of the primary DNA is lost per unit time, and that 0.001 of the secondary DNA becomes primary DNA in the same period of time. The length of time between duplications is normally distributed between 0 and 20 units of time, and the amount of duplication is randomly distributed between 1 and 2 (i.e., none and doubling). With these numbers an average lineage will maintain its DNA content.

Figures 5B to D show three hypothetical lineages generated by this model. Figure 5B would be a lineage that had much DNA in the past, but today is nearly back to where it started. Figure 5C has been losing DNA and may become extinct, and Figure 5D continues to have more and more DNA. These three curves are three of the first four the computer generated. It is easy to see that chance events alone can produce lineages with very different patterns; if they were real organisms, they would have very different evolutionary histories. The requirements are that duplications occur and that some survive. DNA loss and change will follow. The lower curve of Figure 5B shows what would happen to

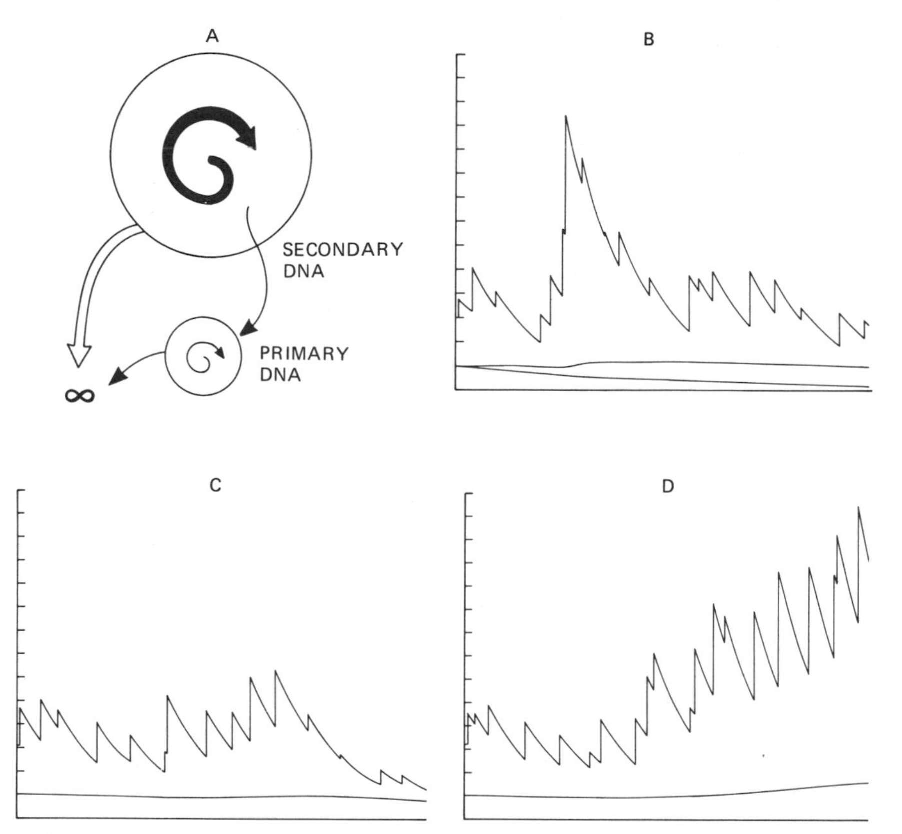

FIGURE 5. A: Diagram of the relationship and relative amounts of primary and secondary DNA. Arrows inside circles indicate nucleotide sequence changes. The two arrows pointing toward infinity indicate the relative amounts of DNA that are lost, and the arrow going from secondary to primary represents new nucleotide sequences that become important to an organism. B to C and D: Jagged graphs show what the evolution of DNA content might look like in three different lineages over long periods of time if increases occured randomly. The abscissa should be considered to be about 600 MY long ±300 MY. The lower curve in (B) shows how the primary DNA amount would decay in the absence of new sequences. The other curve in (B) and the lower curve in (C) and (D) represent the amount of primary DNA under the conditions described in the text.

primary DNA content under the above conditions if there were no additions to it. The other curve at the bottom of this graph and the one at the bottom of the other two graphs shows what the primary DNA looks like with both the loss and the slow addition of new sequences coming from the secondary DNA. (All three lineages started with two units of DNA.)

CONCLUSIONS

There seem to be three basic patterns of DNA increases and decreases in the evolution of higher eukaryotic organisms:

1. The first occurs in organisms with genetic systems that are put together in such a way that DNA increases are not possible or not tolerated. Evolution in lineages of this type is characterized by specialization, loss of parts, and loss of DNA. For a while these lineages can be very successful and produce many species, but in time they become extinct. DNA amounts of 0.1 to 3 pg are typical.

2. Other lineages have from time to time experienced modest increases in DNA content by duplication of either a portion of or all their DNA. Evolution following such duplications can lead to the origin of new and markedly different types of organisms. The origin of new parts along with specialization and loss of parts characterizes this type. These organisms may be part of major evolutionary changes. They typically have 1 to 5 pg of DNA.

3. Finally, there are lineages with very large amounts of DNA, probably the consequence of numerous duplications. These organisms may survive because of whatever virtues bulk amounts of DNA contribute. Large cells and slower rates of cell division and metabolism are typical of very high DNA organisms. The surviving species in lineages that were dominant in the past often are of this type. Specialization may also occur in this group, but it cannot be well correlated with changes in DNA content. DNA amounts range from 10 to more than 100 pg.

Nucleotide sequences in the DNA of eukaryotes fall into either of two partially overlapping types. There are sequences that are involved in protein synthesis, its control, and in any other activities that immediately affect an individual organism's survival in its normal habitat. This is the primary DNA; it may represent only a small amount of an organism's total DNA content. Most DNA nucleotide sequences are probably not essentially involved in the organism's existence. This is the secondary DNA. Nucleotide sequences here are only weakly affected by selection and changes, and losses are more rapid than in the primary DNA.

SUGGESTED READINGS

The stages that can be recognized in the evolutionary history of a lineage through the radiation, plateau, and postplateau periods, with its eventual extinction, are illustrated by B. Rensch, *Evolution above the Species Level,* Columbia University Press, New York, 1959. An interesting computer simulation of phylogenetic history has been made by D. M. Raup, S. J. Gould, T. J. M. Schopf, and D. S. Simberloff, Stochastic models of phylogeny and the evolution of diversity, *J. Geol.* **81,** 525–542, 1973.

The classic paper on the evolution of DNA content is A. E. Mirsky and H. Ris, The deoxyribonucleic acid content of animal cells and its evolutionary significance, *J. General Physiol.*, **34,** 451–462, 1951. Reviews of the DNA contents of vertebrates and plants, respectively, are K. Bachmann, D. B. Goin, and C. J. Goin, Nuclear DNA amounts in vertebrates; and H. Rees, DNA in higher plants. Both these articles are in H. H. Smith *et al.* (eds.), *Evolution of Genetic Systems, Brookhaven Symp. Biol.*, **23,** 1974. Two papers by R. Hinegardner dealing with the DNA content of some invertebrates are Cellular DNA content of the mollusca, and Cellular DNA content of the echinodermata, in *Comp. Biochem. Physiol.*, **47A,** 447–460, 1974, and **49B,** 219–226, 1974.

CHAPTER 12

EVOLUTION OF REPETITIVE AND NONREPETITIVE DNA

GLENN A. GALAU
MARGARET E. CHAMBERLIN
BARBARA R. HOUGH
ROY J. BRITTEN
ERIC H. DAVIDSON

The structure and organization of repetitive DNA sequences has been studied in a number of different animals. Usually a majority of the repetitive DNA consists of short sequences a few hundred nucleotides long which are interspersed with nonrepetitive DNA sequence. The remainder of the repetitive DNA is arranged in long stretches uninterrupted by nonrepetitive sequence. The long repetitive sequence class includes satellite DNA and other well-known multigene families, such as the ribosomal RNAs and the histone genes, as well as repetitive structural genes not yet characterized. We will review some of the experimental evidence on which is based the current knowledge of the characteristics of repetitive DNAs in animal genomes.

Repetition and interspersion frequency of repetitive and nonrepetitive DNA sequence

Table I summarizes some characteristics of 12 animal DNAs in which the repetitive DNA sequence content and sequence organization have been studied. Renaturation kinetics shows that the DNAs of all 12 organisms include repetitive sequences present 10 to 200 times per genome. In addition, most genomes listed possess repetitive frequency classes in which each sequence appears about 1,000 to 5,000 times per genome. Occasionally very simple sequence DNAs present on the order of a million times per genome can also be detected. Other repetition classes may, of course, be present, and to some extent the frequencies listed are only averages, representing components present in various degrees of repetition.

A large fraction of the repetitive sequences are interspersed with nonrepetitive DNA in most of the animals examined so far (Davidson *et al.,* 1975*a*; Davidson and Britten, 1973). Several different techniques have been employed to demonstrate this fact and to obtain quantitative details of the sequence arrangement. A simple and direct demonstration of interspersion of repetitive and nonrepetitive sequence is obtained when DNA fragments 2,000 to 3,000 nucleotides long are reassociated under conditions in which only repetitive sequences can react (i.e., low *Cot,* concentration of reacting fragments multiplied by time). Fragments containing repetitive sequence duplexes formed during the reaction are isolated from those remaining completely single-stranded by passage over a hydroxyapatite column. This reagent binds all DNA fragments any portion of which is in double-stranded form. Though the fraction of the DNA which is actually in repetitive sequences is typically only about 20 to 40 percent, it is usually found that 80 to 100 percent of the long fragments bind to the hydroxyapatite columns after low *Cot* renaturation. It follows that the repetitive sequence elements occur at least every 2,000 to 3,000 nucleotides, interspersed with single-copy regions.

This type of study has recently been employed to examine the DNA sequence organization of five of the marine invertebrates listed in Table I (Goldberg *et al.,* 1975). More sophisticated hydroxyapatite binding experiments in which binding is measured as a function of fragment length have also been carried out with *Xenopus* (Davidson *et al.,* 1973), sea urchin (Graham *et al.,* 1974), and *Aplysia* (Angerer *et al.,* 1975) DNAs. Table I shows that, with the interesting exception of *Drosophila* (Manning *et al.,* 1975), in all the animal genomes studied more than 70 percent of the single-copy DNA is within 2,000 to 3,500 nucleotides of a short repetitive sequence element. Data such as those summarized in Table I now make it possible to state that the regular alternation of short repetitive and longer nonrepetitive DNA sequences must be a general characteristic of most animal kingdom DNAs.

TABLE I. Approximate parameters of the sequence content and sequence organization for 12 animal DNAs

Organism	Genome size (pg)	Fraction of DNA in non-repetitive sequences	Repetitive frequency classes (copies per genome)				Fraction of total repetitive DNA in 200–400 nucleotide long sequences	Fraction of single-copy DNA interspersed with repetitive DNA in 2,000–3,500 nucleotide fragments
			$< 10^3$	10^3–10^4	10^4–16^6	$> 10^6$		
Spisula (clam)	1.2	0.75	30	...	.....	.....	0.60	> 0.70
Crassostrea (oyster)	0.69	0.60	40	3,700	.....	.....	0.35	> 0.75
Aplysia (sea hare)	1.8	0.55	85	4,600	.....	7,000,000	0.60	> 0.80
Loligo (squid)	2.8	0.75	100	4,100	230,000	.....	0.60	> 0.85
Limulus (horseshoe crab)	2.8	0.70	50	2,000	.....	.....	0.75	> 0.70
Cerebratulus (nemertean worm)	1.4	0.60	40	1,200	.....	.....	0.55	> 0.70
Aurelia (jellyfish)	0.73	0.70	180	....	.....	.....	0.60	> 0.80
Strongylocentrotus (sea urchin)	0.89	0.75	100	1,500	.....	.....	0.75	> 0.70
Xenopus (clawed toad)	2.7	0.75	100	2,100	290,000	.....	0.75	> 0.70
Rattus (rat)	3.2	0.75	70	2,000	.....	.....	?	> 0.65
Bos (cow)	3.2	0.65	?	?	60,000	1,000,000	0.55	> 0.65
Drosophila (fruit fly)	0.12	0.75	35	....	.....	.....	0.10	none observed

Repetitive sequence length

Recent evidence shows that both long (greater than 2,000 nucleotides) and short (about 300 nucleotides) repetitive sequences are present in animal DNAs. Length distributions for the repetitive sequence elements have been measured by several techniques. Perhaps the most direct is visualization in the electron microscope of interspersed repetitive sequence duplexes present in fragments significantly longer than the repeated sequence elements (Chamberlin *et al.,* 1975; Manning *et al.,* 1975). The average length of the repetitive DNA sequences may also be obtained by melting reassociated long fragments which have been reacted to low *Cot* and then measuring the hyperchromic shift due to the repetitive duplexes (Davidson *et al.,* 1973; Graham *et al.,* 1974). Another method depends on the use of single-strand specific nucleases, such as the S1 nuclease of *Aspergillus* (Vogt, 1973) and Exo-VII of *Escherichia coli* (Chase and Richardson, 1974*a,b*). While S1 nuclease has both endonucleolytic and exonucleolytic activity, Exo-VII digests single-strand DNA from the ends only. This obviates the possibility of nicking a repetitive duplex. Repetitive sequence length is determined with these enzymes by reacting DNA fragments several thousand nucleotides long at low *Cot* and then digesting away the fraction remaining single-stranded. The length distribution of the enzyme-resistant duplex provides an estimate of the lengths of the repetitive sequences. Data obtained by all these methods agree in that sizable fractions of the repetitive DNA (again with the exception of *Drosophila*) exist as 200 to 400 nucleotide sequences. The electron-microscope approach has been used in the case of *Drosophila* and *Xenopus* DNA, as noted above, and S1 and Exo-VII nuclease as well as electron microscopy has been applied to sea urchin DNA. Most of the remaining data listed in Table I were obtained by the S1 nuclease and hyperchromicity methods. In all species studied the amount of short repetitive DNA sequence is more than sufficient to account for the amount of interspersed nonrepetitive DNA observed in that genome. The nonrepetitive sequence generally extends for at least 1,000 to 2,000 nucleotide before terminating in repetitive sequence regions. The remaining repetitive DNA is present in longer sequences, the length of which is unknown, except it is greater than 2,000 nucleotides.

Comparison of long and short repetitive DNA sequence classes

Neither long nor short repetitive sequences are restricted to one or a few repetitive frequency classes. This has been shown in several ways:

Long and short S1 resistant repetitive sequences have been isolated directly in sea urchin DNA. Here agarose columns are used to partially separate S1 nuclease resistant repetitive duplexes of various lengths. When they are reacted with excess whole DNA, it is found that all the degrees of repetition present in unfractionated total repetitive DNA appear in both the long and short repeat fractions. Furthermore, self-reaction kinetics shows that both fractions have about the same complexity. Similar data exist for *Xenopus* DNA. These results exclude the possibility that the long repetitive sequences consist only of large blocks of satellite-like DNA.

Are sequences held in common in the long and short repeat classes? Such an organization would explain the presence of both long and short repetitive sequences in all frequency classes in the genome. No clear answer can be given to this question as yet, but the following preliminary experiment carried out with sea urchin DNA can be described. The isolated long and short sequence classes were reassociated separately in the presence of excess amounts of DNA of the other length class. While all the short repetitive tracer reacts with excess long repetitive DNA, it does so at a rate about 10 times slower than the long driver self-reassociation. This indicates that only a small fraction of the mass of the long repetitive driver contains all the sequences present in the short repeats. It is still unclear if the cross-reacting components represent contaminants due to incomplete purification of the length classes or if there are indeed minor fractions of each length class which contain all the sequences in the other length class. Nonetheless the bulk of the repetitive DNA sequence in each length class must occur rarely (if at all) in the other length class.

It has been known for some years that extensive divergence exists within each family of repetitive sequence (Britten and Kohne, 1968). When S1 resistant repetitive duplexes are chromatographed on agarose columns and fractions containing different length duplexes are melted, the melting temperature of the duplexes declines sharply with decreasing sequence element length (Davidson *et al.*, 1973; Britten and Davidson, 1975; Goldberg *et al.*, 1975). Part of this decrease is due merely to the effect of duplex length on melting temperature; but even when this small correction is included, the correlation between repetitive sequence length and divergence remains striking. Table II summarizes such studies, which have so far been carried out with the DNA of four organisms. In all cases the long repetitive duplexes melt with a corrected average temperature no lower than 1°C below long native DNA. Because a decrease in melting temperature of 1°C is equal to about 1 to 1.5 percent mismatch in base sequence (Laird *et al.*, 1969; Britten *et al.*, 1974), less than 2 percent divergence in base sequence exists within each long repetitive sequence family. On the other hand, the 200 to 400 nucleotide long repetitive sequence class melts 8 to 12°C below native DNA when corrected for duplex length, indicating that the various copies of these sequences differ an average of at least 10 to 15 percent in base sequence. It should be noted that this large difference in the melting characteristics of long and short repetitive DNA cannot be an artifact of the enzyme digestion used in the preparation of these fractions. This has been shown directly by isolating the high- and low-melting DNA fractions formed in the low *Cot* reassociation of 2,000 nucleotide long fragments and measuring their nonrepetitive sequence content. The low-melting fraction contained a large amount of non-repetitive sequence (55 percent), as should be the case for the short interspersed sequence fraction. In contrast, and as expected for long repetitive sequence, the high-melting fraction included very little non-repetitive DNA (Britten and Davidson, 1975).

An additional aspect of DNA sequence organization relevant to repetitive DNA evolution is the question of local interspersed repetitive sequence homology. Direct visualization of long reassociated *Xenopus* DNA in the electron microscope failed to detect a significant number of the structures which would be predicted if the adjacent repetitive repeats are frequently the same sequence (Chamberlin *et al.,* 1975). Observation of Exo-VII treated reassociated products has now extended such measurements to longer fragment lengths, but as yet no indication of local homology among interspersed sequences has been found. Such experiments cannot efficiently reveal homology over very long regions, however, e.g., every three or four repeats. The tentative conclusion is that the interspersed repetitive sequences within a given region of DNA tend to be dissimilar in sequence.

TABLE II. Thermal stabilities of long and short repetitive sequences in the DNAs of four organisms. (The values given are the melting temperatures of duplexes relative to long native DNA, corrected for duplex length)

Organism	Long repeats, $\geq$ 1,500 nucleotides (°C)	Short repeats, 300 nucleotides (°C)	Source
Spisula (clam)	−1	−8	Goldberg *et al.,* 1975
Strongylocentrotus (sea urchin)	<−1	−9	Britten, Graham, Eden, & Davidson, unpublished
Xenopus (clawed toad)	−1	−11.5	Davidson *et al.,* 1973
Bos (cow)	+3.3	−11.5	Britten, unpublished

In summary, there appear to be two general types of repetitive sequences. One class consists of short sequences of a length of about 300 nucleotides. In most of the genomes which have been studied the bulk of these are interspersed with nonrepetitive sequences. These short repetitive sequence elements occur in all frequency classes, except perhaps in some of the most rapidly reassociating very simple sequence DNAs. The related copies of each of these sequences have substantially diverged within each genome so that the products of their reassociation average 10 to 15 percent mismatch. A second class of repetitive DNA is found in sequence elements exceeding 2,000 nucleotides in length. These long repetitive sequences are also represented in all repetitive frequency classes. The various copies of sequences occurring in the long repetitive regions differ only very slightly from each other, displaying less than 2 percent mismatch after renaturation. The sequences in each length class occur rarely, if at all, in the other length class.

Function and evolution of long repetitive DNA sequences

The long repetitive DNA sequences comprise 5 to 15 percent of the total genomic DNA in many animal genomes, or about 0.6×10^8 to 3×10^8 nucleotides (Table I). It is possible that much of this DNA is arranged in long stretches of tandemly repeated sequence. Satellite DNAs, which can account for a significant fraction of eukaryotic DNA, are certainly included in these long repeats. Also included are the transcribed multigene families now fairly well characterized in *Xenopus*. The genes coding for high-molecular-weight rRNA, 5S rRNA, and transfer RNAs constitute 2 to 3 percent of the total *Xenopus* DNA (Brown *et al.*, 1971; Speirs and Birnstiel, 1974; Dawid *et al.*, 1970; Clarkson *et al.*, 1973). We also know that some of the long repeats are multigene families coding for mRNAs. In the sea urchin, for example, approximately 0.5 to 0.8 percent of the genome is in the histone mRNA multigene family (Birnstiel *et al.*, 1974).

There is reason to believe that there are other repetitive structural genes yet to be discovered (Hood *et al.*, 1975). A substantial fraction of polyadenylated mRNA is complementary to repetitive DNA in mammalian cell lines. (See review in Lewin, 1975.) These mRNAs consist entirely of transcripts from repetitive gene sequences, and their length is aproximately equal to that of the rest of the mRNA (Klein *et al.*, 1974; Campo and Bishop, 1974). Their DNA templates would belong to the long repetitive sequence class. Preliminary indications (e.g., Ryffel and McCarthy, 1975) are that at least 2 percent of the total DNA mass is included in repetitive DNA sequence represented in polyadenylated mRNA in mammalian cell lines. This is clearly a minimum estimate of the translated long repetitive sequences because it excludes nonpolyadenylated mRNAs, mRNAs not transcribed in these particular cell lines, and any nontranscribed spacer sequence present in multigene mRNA families. It is not known, as yet, if these repetitive structural gene sequences are in general clustered, as is the case for the histone genes. We conclude that in a genome such as that of *Xenopus,* where about 6 percent of the total DNA is in long repetitive sequences, most of this sequence class may be accounted for as transcribed multigene families.

The evolution of multigene systems is now under intensive investigation and has been reviewed elsewhere (Smith, 1974; Southern, 1975; Hood *et al.,* 1975; Brown and Sugimoto, 1974). The means by which evolution of multigene sequences occurs seem to be a direct consequence of their tandem arrangement. At present the most widely accepted mechanism is unequal crossover, i.e., out-of-register pairing and recombination within long stretches of tandem repeats. This mechanism provides an explanation for the paradoxical observation that the sequences of multigene families (for example, rRNA spacer) diverge rapidly from species to species while maintaining near homogeneity within the genomes of each species. In a more general sense the high-melting behavior of renatured long repetitive sequences means that these sequence families are either under tight selective control (for example, rRNA or histone genes) or, considered as a set, have been assembled recently in evolution (e.g., the species-specific spacers). The postulate of unequal crossover provides a means for the generation of new sets of repetitive sequence, that is, if one assumes a tandem arrangement. This, however, is not the only way in which new long repetitive regions might come into existence. Until more is known about the nature of the long repetitive sequence class, their history must be considered a subject for speculation.

Function and evolution of short repetitive DNA sequences

The function of the short repetitive DNA sequences is not as yet understood. However, the nonrandom distribution of interspersed repetitive sequence lengths and their spacing in nonrepetitive DNA shows that the interspersed pattern of sequence organization has been selected. Therefore the short repetitive sequences must play some role in the life of the organism. Furthermore, it has recently been found that the structural genes active in sea urchin embryogenesis are located in the immediate vicinity of interspersed repetitive sequences (Davidson *et al.,* 1975*b*). Though this suggests some function, perhaps regulatory (Britten and Davidson, 1969), for these repetitive sequence elements, such a function cannot pertain to more than a minor fraction of the interspersed repetitive sequence in the genome. It now seems likely that less than 10 percent of the single-copy genome in the sea urchin actually consists of structural genes (e.g., see Britten and Davidson, 1975) and to a crude approximation this would also be the maximum fraction of interspersed repetitive sequences likely to be adjacent to structural genes.

A substantial fraction of the repetitive sequence elements are also transcribed in heterogeneous nuclear RNA (hnRNA). This RNA species has been shown to include interspersed repetitive and nonrepetitive sequences on the same RNA molecule (Darnell and Balint, 1970; Holmes and Bonner, 1974; Smith *et al.,* 1974; Molloy *et al.,* 1974). Thus the organization of hnRNA is quite similar to that of the total DNA. Because as much as 30 percent of the sea urchin single-copy DNA is transcribed into nuclear RNA at the gastrula stage alone (Hough *et al.,* 1975), it is clear that a large fraction of the interspersed repeats must be represented in the hnRNA. Indeed this has been found to be the case in ascites nuclear RNA (Holmes *et al.,* 1974) and according to preliminary experiments in the sea urchin material as well. What function these sequences serve remains obscure because the role of hnRNA itself is still unclear. In some hypotheses these transcribed repetitive elements have been assigned a regulatory role, either as diffusable signals active in the control of transcription (Britten and Davidson, 1969) or as signals required for the transcription or processing of the long hnRNA transcripts (Georgiev *et al.,* 1974; Jelinek *et al.,* 1974). RNA complementary to repetitive DNA is also present in *Xenopus* oocytes, and the melting characteristics of the hybrids again suggest that the repetitive DNA belongs to the short repeat class (Hough and Davidson, 1972).

By comparing various rodent DNAs it has been found that the high thermal stability fraction of the repetitive sequence in each genome tends to be species-specific while the lower thermal stability fraction is more extensively shared between related species (Rice, 1971, 1972). Among the species for which a typical interspersed sequence arrangement has been demonstrated is the rat (Table I) studied by Bonner *et al.* (1974). From the data reviewed above (Table II) we may associate the high thermal stability fraction with the long repetitive sequences and the lower thermal stability fraction with the short interspersed sequences. We conclude that at least some of the short repetitive sequence families present in a given genome tend to be older, evolutionarily speaking, than are most of the long repetitive sequence sets.

In *Xenopus laevis* 75 percent of the repetitive DNA is included in the 300-nucleotide interspersed sequence class. This amounts to about 18 percent of the genome, or some 4.9×10^8 nucleotides. Therefore, the genome contains some 1.6×10^6 300-nucleotide repetitive sequence elements. The major frequency of repetition averages 2,000 copies of each sequence (Davidson *et al.,* 1973); so there are in this repetitive class about 800 distinct repetitive sequence families. In addition a complex class of long repetitive sequences exists. In the following section we consider some experimental evidence on the evolution of this huge library of repetitive sequences. Our approach has been to measure the changes which have occurred in the repetitive sequence components since the divergence of *Xenopus laevis* from one of its closer congeners, *X. borealis* (formerly *X. laevis borealis;* see Tymowska and Fischberg, 1973). These changes have affected the nucleotide sequences themselves and the frequency with which related repetitive sequences occur in the two genomes.

DIVERGENCE AND HOMOLOGY IN THE REPETITIVE DNAs OF *XENOPUS LAEVIS* AND *X. BOREALIS*

Reassociation kinetics of *X. laevis* and *X. borealis* DNAs

The rates of reassociation of the various classes of repetitive sequence in *X. laevis* DNA have been studied previously (summary in Davidson *et al.,* 1973). Figure 1 displays the reassociation of whole *X. laevis* DNA sheared to 450 nucleotides (solid circles, solid line) as well as the various kinetic components (dashed lines, curves A to D). While these components are not unique choices and may in some cases represent averages rather than discrete frequency classes, they portray the best least-squares fits to the data if one assumes a small number of components. The major repetitive sequence class in which each sequence occurs about 2,000 times (curve *B*) is present on about 31 percent of the 450 nucleotide fragments. In addition there are smaller repetitive components in which the sequences are present an average of 100 times (curve *C*) and over 10,000 times (curve *A*). The most slowly renaturing component is the single-copy sequence (curve *D*). The actual amount of single-copy sequence in the genome is of course underestimated in an experiment such as that in Figure 1 because of sequence interspersion. About a third of the total single-copy sequence is included in fragments which also contain repetitive sequence elements and hence bind at lower *Cot*s than do fragments containing only single-copy sequence.

The reassociation profile for *X. borealis* DNA (open circles, dotted line) is remarkably similar to that of *X. laevis* DNA. In fact the *X. borealis* data can easily be fit if one assumes exactly the same rates of reaction for the *X. borealis* kinetic components as for the *X. laevis* genome. When this is done, the RMS (root-mean-square) error of the fit remains the same as if the components are left free. As Figure 1

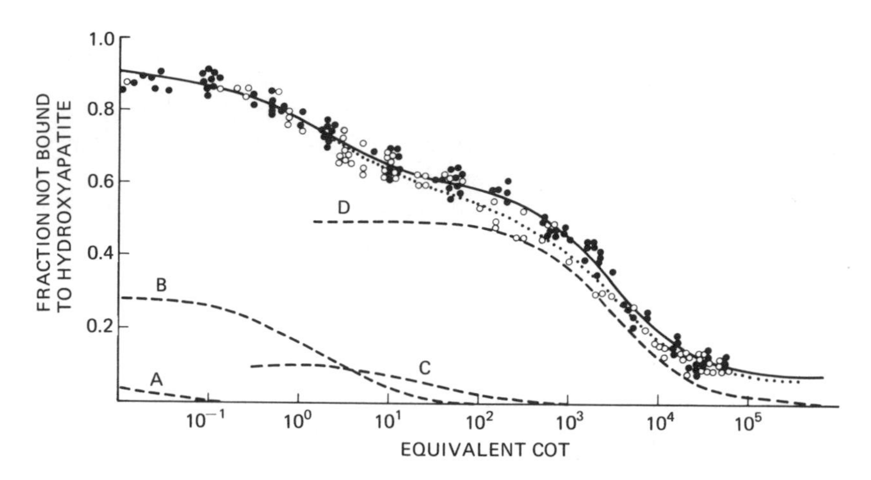

FIGURE 1. Reassociation of 450-nucleotide-long X. *laevis* and X. *borealis* DNAs. The solid and dotted curves represent the best least-squares solutions obtained with a computer if one assumes second-order kinetic components (Britten *et al.*, 1974). The components shown are those calculated for the X. *laevis* reaction (-•—•-), and the same rates were used to obtain the fit shown for X. *borealis* DNA (·○···○·). (Additional experimental details are given at the end of the chapter.)

shows, only slight or imperceptible differences may exist in the quantity of each repetitive sequence class in these two congeners.

Isolation of repetitive DNA fractions from the two *Xenopus* genomes

The isolation and characterization of *X. laevis* repetitive ^{3}H DNA has been described (Hough and Davidson, 1972). Repetitive DNA is prepared by collecting from a hydroxyapatite column those DNA fragments which form duplex structures during incubation to *Cot* 50 (*X. laevis*) or *Cot* 100 (*X. borealis*). This *Cot* is sufficient for the reaction of the small, very fast component (curve A in Figure 1) and the major repetitive class (curve *B* in Figure 1) but not for the entire slow repetitive class (curve *C* in Figure 1). The reaction of this isolated repetitive DNA fraction with whole *X. laevis* DNA is shown in Figure 2. A repetitive *X. borealis* fraction was prepared similarly, and its reaction with whole *X. borealis* DNA is also shown in Figure 2. Sixty to seventy percent of both tracers contain sequences present about 1,000 to 2,000 times per genome, reassociating with second-order kinetics at the rate characteristic of the major repetitive sequence class (0.72 M^{-1} sec^{-1}). The reaction of the small portion of these tracers which belong to the slow repetitive component (curve *C* of Figure 1) is undetectable because of scatter in the terminal portions of the data. Completion of the reaction at 75 to 80 percent of tracer bound is due principally to its short fragment size, which also decreases its thermal stability.

Figure 2 provides additional evidence for the similarity between the repetitive DNAs of *X. laevis* and *X. borealis* as suggested by the overall reassociation profiles in Figure 1. Thus, the isolated repetitive DNA fractions of both genomes renature with their respective whole DNAs at identical rates.

Interspecific reassociation of repetitive DNA fractions

Figure 3A shows the reassociation of *X. laevis* repetitive DNA tracer with an excess of *X. borealis* DNA, and Figure 3B shows the reassociation of *X. borealis* repetitive DNA tracer with an excess of *X. laevis* DNA. These reactions may be compared with the reassociation of the same tracer preparations with an excess of homologous DNA as shown in Figure 2 and reproduced on Figure 3 as dotted curves. All four reactions appear to approach the same degree of completion (about 75 to 80 percent). The relative retardation of at least part of the interspecies

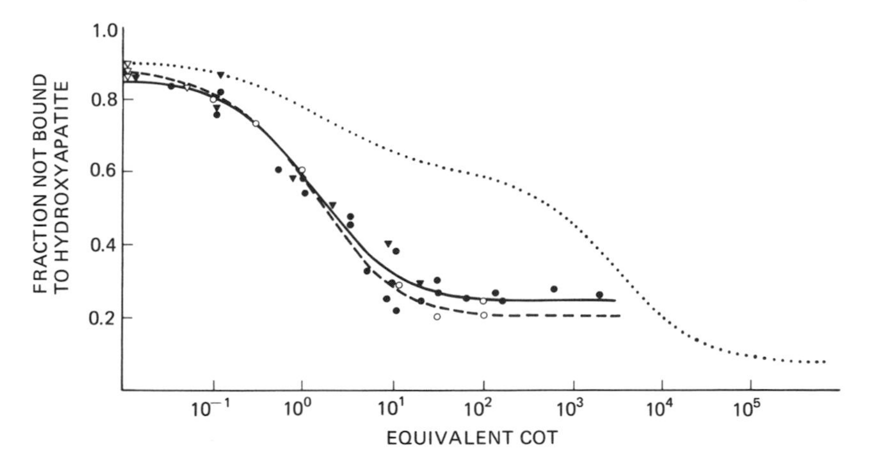

FIGURE 2. Reassociation of isolated repetitive DNA fractions with homologous whole DNA. Reaction of isolated repetitive ^{3}H DNA from *X. laevis* (—•—) and from *X. borealis* (—o—) with their respective whole homologous DNA is shown. Also indicated is the self-reassociation of *X. laevis* (▼) and *X. borealis* (▽) repetitive DNA in the absence of whole DNA driver. The *X. laevis* data are reproduced from Hough and Davidson (1972). [For comparison the reassociation of whole *X. laevis* DNA (from Figure 1) is illustrated as well (dotted line).]

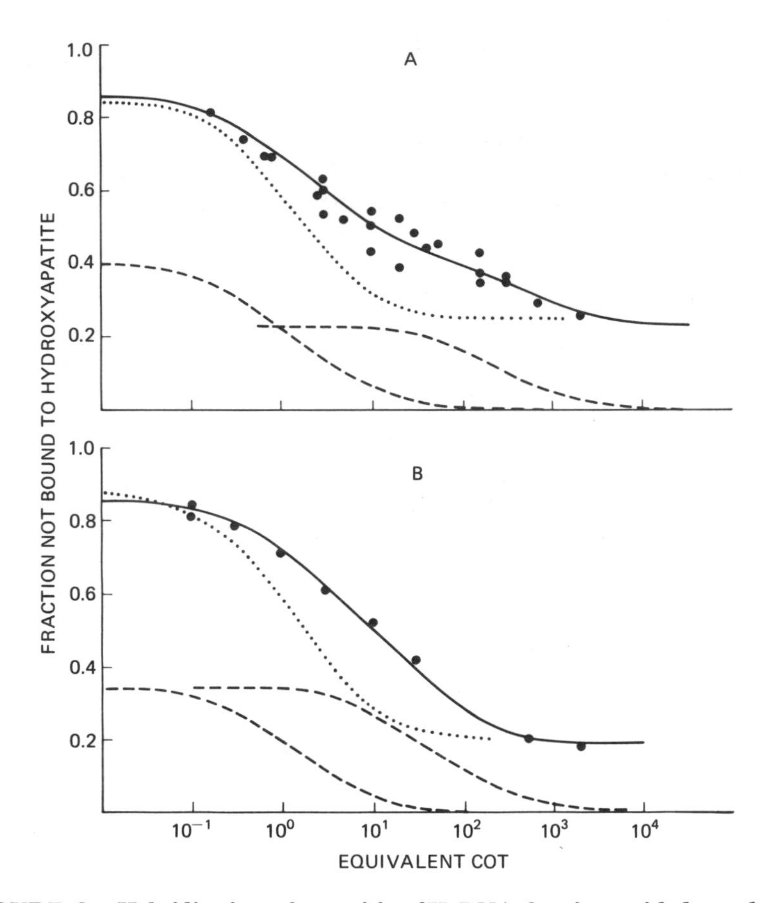

FIGURE 3. Hybridization of repetitive ^{3}H DNA fractions with heterologous whole DNA. A: Hybridization of repetitive X. *laevis* ^{3}H DNA with X. *borealis* DNA. The solid curve through the data is the sum of two second-order kinetic components (indicated by the dashed curves) fit by least-squares methods to the data. (The dotted line is the reaction of the repetitive X. *laevis* ^{3}H DNA fraction with whole X. *laevis* DNA reproduced from Figure 2.) B: Hybridization of repetitive X. *borealis* ^{3}H DNA with whole, unlabeled X. *laevis* DNA. The data are again fit with two second-order components by least-squares methods. These are indicated by the dashed lines, and the solid curve is the summed reaction. (The dotted line is the reassociation of repetitive X. *borealis* ^{3}H DNA with whole X. *borealis* DNA reproduced from Figure 2.)

reactions is evident. From the similar extents of reaction we infer that there are representatives of all repetitive sequence families in the DNA of both species. However, the two sets of repetitive sequences evidently differ in their repetition frequency.

The retardation in the rate of interspecies reaction suggests that there have been some changes in the number of members of some of the sets of repeated sequences. The kinetic analyses shown on Figure 3 are the simplest ways to represent the interspecies reaction kinetics, viz., in terms of two components, one of which has the rate of the homologous reaction and one of which is retarded. This solution suggests that there are components with up to a 100-fold retardation in rate of reassociation. In reality there may be a wide range of different degrees of retardation in the interspecies reactions for different sets of repeated sequences. The thermal stability of the resulting duplexes has to be taken into account in order to derive the frequency of occurrence of repetitive sequences from the rates of reassociation measured in the experiments.

Even at its optimum temperature the renaturation rate for severely mismatched duplexes is somewhat retarded compared with the rate for perfectly matched duplexes (Wetmur and Davidson, 1968; Bonner *et al.*, 1973). Furthermore, the rate begins to approach its maximum value only when the incubation temperature is about 15°C below the melting temperature of the duplexes.

An analysis of the melting characteristics of the hybrids formed by the *X. laevis* repetitive ^{3}H DNA with *X. laevis* DNA and with whole *X. borealis* DNA was therefore carried out, and representative thermal elution profiles are shown for both homologous and heterologous duplexes in Figure 4. The melting temperature of the homologous duplexes is about 3.5°C lower than that of the repetitive driver DNA. It is probable that this difference is due to the shorter mean fragment length of the final tracer preparation. Figures 4C and D show that the heterologous tracer-driver DNA duplexes melt very similarly to the homologous ones. Only a 1 to 2°C depression in melting temperature is noted relative to the homologous tracer duplexes. This very slight temperature reduction is also independent of *Cot* over the range from *Cot* 3 (Figure 4C) to *Cot* 150 (Figure 4D). It is evident that any more divergent heterologous duplexes would not be observed in these experiments because their thermal stability would be too close to the experimental incubation temperature. The result is that this procedure possibly provides an artificially high estimate of the mean heterologous duplex stability.

The differential melting profiles of the *Cot* 150 homologous and heterologous reactions are compared in Figure 4E. Only a small fraction of the observed heterologous duplexes melt at significantly lower temperatures than do the homologous duplexes. Therefore any correction for the effect of divergence on the rate of the heterologous reaction compared with the homologous reaction is likely to be small. We are thus left with a rather surprising conclusion: While most of or all the repetitive sequence families of *X. laevis* are represented in *X. borealis* DNA, many of the sequences are present at 10 to 100 times lower frequencies in the latter genome. Furthermore, the results are complementary; so an equivalent amount of the *X. borealis* repetitive sequences also seems to occur at much lower frequencies in the *X. laevis* genome than it does in the *X. borealis* genome. Two mechanisms could be responsible for the observed evolutionary decrease in repetitive sequence frequency. These are (1) divergence of a large portion of the repetitive sequences in some families so that heterologous duplexes are no longer stable under our conditions and (2) reduction in copy number of some sequences as a result of unequal crossover and selection. The first mechanism can apply either to interspersed or to long repetitive sequences, while the second must refer mainly to long tandemly repeated regions. The present data do not permit us to decide which of these mechanisms has been more important in the evolution of the *Xenopus* genome.

INTERSPECIFIC REACTION OF NONREPETITIVE DNA SEQUENCES

To provide a standard of comparison for the results obtained with the repetitive sequence fractions, the homology between *X. laevis* and *X. borealis* nonrepetitive DNAs has also been measured. A mainly nonrepetitive fraction was isolated from *X. laevis* DNA by passing the DNA over hydroxyapatite after incubation to *Cot* 2,500. The DNA which failed to bind to the column was harvested. The reaction of this nonrepetitive *X. laevis* tracer with excess whole homologous *X. laevis* DNA is shown in Figure 5A. About 75 percent of the tracer reassociates as single-copy sequence. In addition about 10 percent may be complementary to the slow repetitive component in whole DNA, the sequences of which are present about 100 times per genome. (See curve C of Figure 1.) The nonrepetitive *X. laevis* ^{3}H DNA was reacted with a large excess of whole *X. borealis* DNA in 0.41-*M* phosphate buffer at 60°C (equivalent to about 53°C in 0.12-*M* phosphate buffer). The results of this experiment are shown in Figure 5B. Duplex formation was measured as usual by binding to hydroxyapatite in 0.12-*M* phosphate at 60°C. From this reaction we conclude that most of the nonrepetitive sequence is shared between these two species. The arguments leading to this conclusion are now considered.

Within the range of *Cot*s technically accessible, the reaction of the nonrepetitive *X. laevis* tracer with *X. borealis* DNA is incomplete. Only about 40 percent of the tracer reacts by *Cot* 50,000 (Figure 5B). The melting profiles of the heterologous duplexes and of the homologous tracer-driver duplexes are shown in Figure 6. Duplexes formed between *X. laevis* nonrepetitive tracer and homologous whole DNA melt at about 77°C (Figure 6A). For comparison this is about 4°C below the melting temperature of single-copy DNA duplexes, 81.5°C (dashed line,

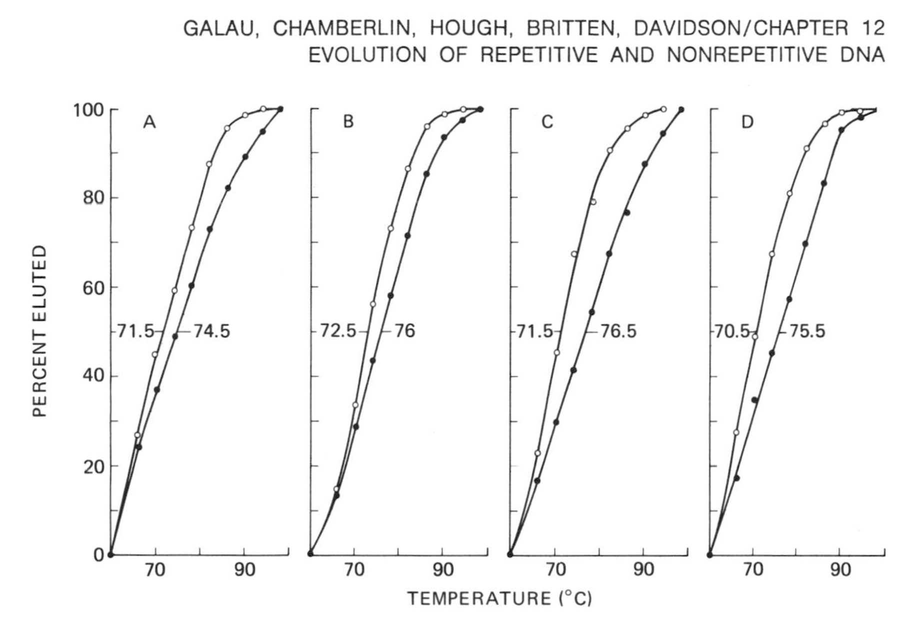

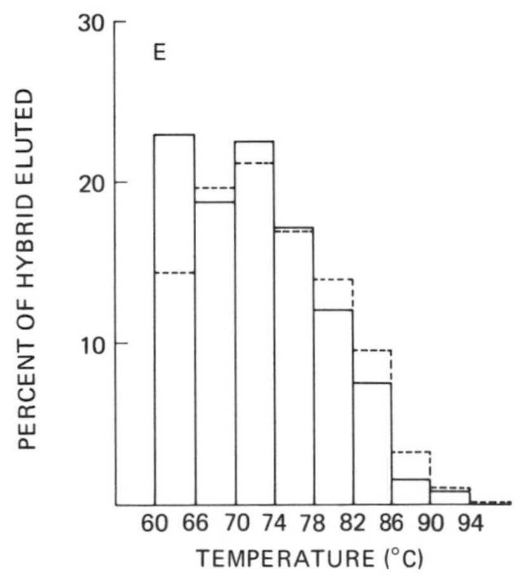

FIGURE 4. Representative melts of repetitive X. *laevis* ^{3}H DNA hybrids. Repetitive X. *laevis* ^{3}H DNA was incubated with whole X *laevis* or X. *borealis* DNA using the same methods as for Figures 2 and 3. The numbers near each curve indicate the temperature at which one-half of the DNA has eluted (T_m). A, B: Thermal elution of repetitive X. *laevis* ^{3}H DNA duplexes with driver X. *laevis* DNA after incubation to *Cots* 3 and 150, respectively. The elution profiles of the repetitive ^{3}H DNA duplexes (—○—) and the driver X. *laevis* DNA (—●—) duplexes are shown C, D: Thermal elution of repetitive X. *laevis* ^{3}H DNA duplex with driver X. *borealis* DNA after incubation to *Cots* 3 and 150, respectively. The elution profiles of the repetitive ^{3}H DNA hybrids (—○—) and the driver X. *borealis* DNA duplexes (—●—) are shown. E: Comparison of the thermal stability of homologous and heterologous repetitive ^{3}H DNA duplexes. The thermal elution profile of repetitive X. *laevis* ^{3}H DNA hybrids formed at *Cot* 150 with excess X. *laevis* (dashed line) and with X. *borealis* DNA (solid line) are replotted incrementally from (B) and (D). The graph shows the percent of the total material which elutes in each temperature interval. Within experimental error the two driver DNAs melted similarly. The elution profiles of the unlabeled driver DNAs are not shown. Similar results are obtained when *Cot* 20 rather than *Cot* 150 duplexes are compared.

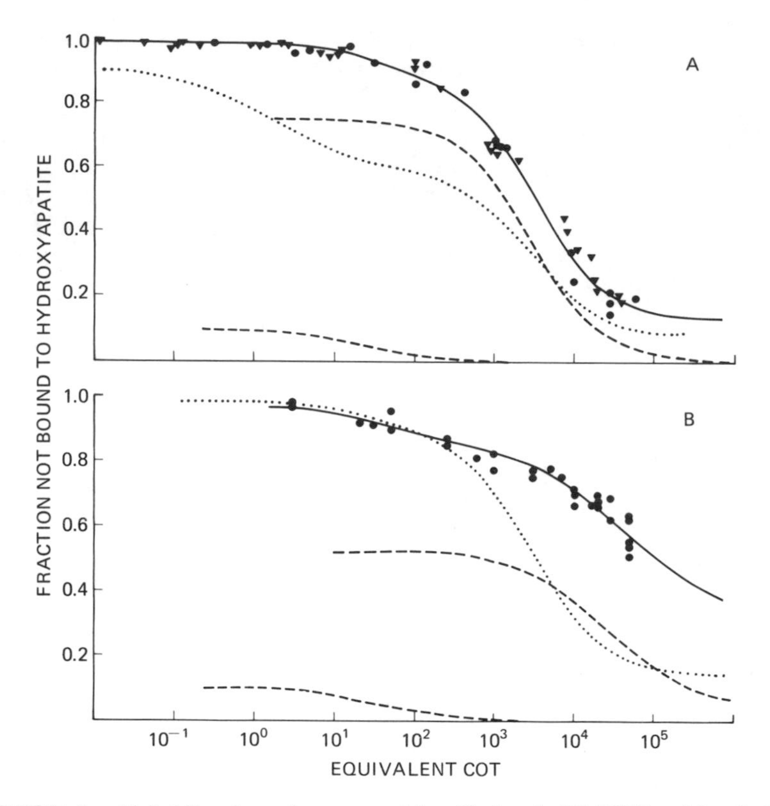

FIGURE 5. Hybridization of nonrepetitive X. *laevis* ^{3}H DNA with whole X. *laevis* and X. *borealis* DNA. A: Reassociation of nonrepetitive X. *laevis* ^{3}H DNA with whole X. *laevis* DNA. Also indicated are self-reassociation data for nonrepetitive X. *laevis* DNA (▼) in the absence of whole driver DNA. The data are reproduced from Hough and Davidson (1972). The solid curve is the result of a least-squares fit including two second-order components, indicated by the dashed lines. This particular solution is the result of fixing the rate of the main component at 0.00035 M^{-1} sec^{-1} (the rate of the single-copy DNA in whole DNA from Figure 1). [Included for comparison is the reassociation of whole X. *laevis* DNA (dotted line) reproduced from Figure 1.] B: Hybridization of nonrepetitive X. *laevis* ^{3}H DNA with whole X. *borealis* DNA. The solid curve is the result of a fit of two components to the data (indicated by the dashed lines). [Included for comparison is the reassociation of nonrepetitive X. *laevis* ^{3}H DNA with unlabeled X. *laevis* DNA (dotted line) reproduced from (A).]

Figure 6A). Figures 6B to D show the melting profiles of the heterologous tracer-driver duplex fractions at *Cot*s 3,000, 16,000, and 30,000, respectively. In all three samples, 50 percent of the tracer becomes single-stranded at 67°C, 10°C lower than the homologous tracer-driver duplexes. The shape of the heterologous melts shows that a substantial fraction of the hybrid may have formed duplexes at the 53°C incubation criterion but was melted at 60°C when loaded on the hydroxyapatite columns. Even for a duplex population melting at 67°C the rate of hybridization of tracer with the carrier should be only about 0.28 of that of the self-reassociation of the driver DNA under the conditions used (Bonner *et al.,* 1973). The appropriate mathematical form for the reduction of such heterologous kinetic data is shown in the section on Experimental Details. When the rate reduction of 0.28 times the homologous rate is included, the calculation shows that the reaction would have terminated with 60 to 70 percent of the heterologous tracer in duplex. However, it can be argued that the mean melting temperature would have been only 64°C had the thermal stability measurement been made from 53°C rather than from 60°C. On this basis (i.e., because the

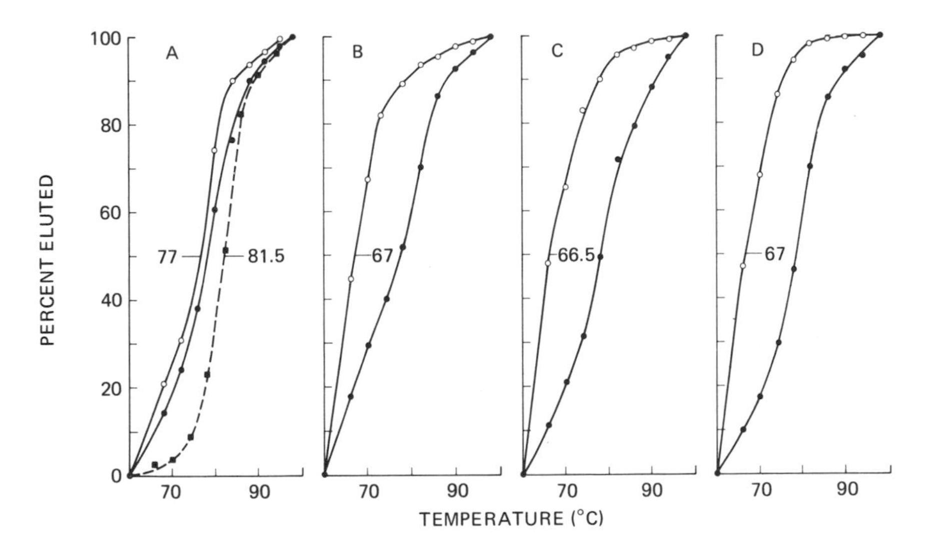

FIGURE 6. Representative melts of nonrepetitive X. *laevis* ³H DNA duplexes. A: Thermal elution of nonrepetitive X. *laevis* ³H DNA duplexes with driver X. *laevis* DNA formed at *Cot* 10,000. The elution profiles of nonrepetitive ³H duplexes (—○—) and driver X. *laevis* DNA duplexes (—●—) are shown [Included for comparison is the elution from a parallel column of unlabeled nonrepetitive X. *laevis* DNA duplexes (—■—).] B to D: Thermal elution of nonrepetitive X. *laevis* ³H DNA hybrids with driver X. *borealis* DNA after incubation of *Cots* 3,000, 16,000, and 30,000, respectively. The elution profiles of nonrepetitive ³H DNA duplexes (—○—) and driver X. *borealis* DNA duplexes (—●—) are shown.

reactions were carried out at 53°C criterion), the curve shown in Figure 5B (solid line) is calculated for the reassociation of DNAs so divergent in sequence as to melt (50 percent) at 64°C. The renaturation rate for such a DNA fraction would be approximately 0.2 times that of the single-copy sequences in the driver DNA (Bonner *et al.*, 1973), and it can be seen that with this assumption almost complete heterologous cross reaction of those sequences able to react with homologous DNA would have been attained at completion.

Further experiments are shown in Table III. The *X. laevis* nonrepetitive tracer was reacted to *Cot* 30,000 with *X. borealis* DNA (reaction 1), and the hybridized tracer and unhybridized tracer were separately recovered and rereacted with *X. laevis* and *X. borealis* DNA. Rereaction of the *X. laevis* tracer with *X. laevis* driver DNA provides a control for the fraction of the tracer which has survived the long incubations and chromatography steps and remains reactable. Thus the tracer which reacted with the *X. borealis* DNA is still over 80 percent reactable (reaction 2). However the fraction which failed to react with *X. borealis* DNA by *Cot* 30,000 remains only about 52 percent reactable when assayed at 60°C (reaction 4) and 73 percent reactable when assayed at 50°C (reaction 6). If the fraction of the *X. laevis* tracer which reacted with *X. borealis* DNA was a selected fraction of more homol-

TABLE III. Reassociation of nonrepetitive X. *laevis* ^{3}H DNA with X. *laevis* and X. *borealis* DNA

Selected nonrepetitive X. *laevis* ^{3}H DNA tracer	Reaction	Excess DNA driver*	Temperature of incubation and assay† (°C)	Fraction of selected tracer in hybrid at *Cot* 30,000
Starting unfractionated tracer	1	*X. borealis*	60	0.38
Tracer in heterologous hybrid at *Cot* 30,000	2	*X. laevis*		0.82, 0.85 (av. 0.83)
	3	*X. borealis*		0.54, 0.77 (av. 0.66)
Tracer not in heterologous hybrid at *Cot* 30,000	4	*X. laevis*	60	0.51, 0.51, 0.54 (av. 0.52)
	5	*X. borealis*		0.16, 0.18, 0.25 (av. 0.20)
	6	*X. laevis*	50	0.73
	7	*X. borealis*		0.32, 0.37 (av. 0.34)

*Excess DNA was added to the fractionated tracer to 50,000 times the tracer concentration and 20 to 200 times the residual fractionated *X. borealis* carrier DNA.

†Incubation was in 0.41-*M* phosphate buffer, and the hydroxyapatite assay was conducted in 0.12-*M* phosphate buffer.

ogous sequences, all the reactable tracer, or over 80 percent of the total tracer, should have been able to rereact with the *X. borealis* DNA. On the other hand if the failure to achieve complete heterologous reaction the first time (i.e., in reaction 1) was merely kinetic, only about 40 percent of the single-copy sequence should again have been able to react on a second exposure to *X. borealis* DNA. Because the starting tracer includes about 10 percent repetitive sequence (Figure 5A), the expected amount of reaction in the latter case is around 68 percent. Reaction 3 shows that the kinetic explanation is probably the correct one, because the average value obtained, 66 percent, is close to prediction. This interpretation is strengthened by the results of reaction 5. Here the fraction of *X. laevis* tracer not reacting with *X. borealis* DNA on first exposure is reassociated once more with *X. borealis* DNA. Again we are faced with two distinct predictions. If the *X. laevis* DNA which initially reacts with *X. borealis* DNA is a special set of sequences, rereaction of the nonbound tracer with *X. borealis* DNA should yield no duplex. If, however, failure of the initial reaction was kinetic, rereaction to *Cot* 30,000 should again yield about 40 percent of the reactable single-copy sequence present. In the latter case we expect about 20 percent total reaction because only half the DNA remains reactable at this stage (shown in reaction 4). This is exactly what is found. When the incubation and assay temperatures are both decreased by 10°C, more tracer reacts with both homologous and heterologous DNAs. However, there is proportionately more reaction with the heterologous DNA (reaction 7). This finding confirms the presence of marginally stable heteroduplexes.

To summarize these experiments, we make the following points: At least 70 percent of and possibly all the single-copy DNA sequences of *X. laevis* are represented in *X. borealis* DNA, and most of the sequences have diverged greatly since the separation of the two species. It can be calculated that the average sequence divergence is at least 10 to 15 percent of the base pairs in the single-copy fraction. Unfortunately the relatedness of all the single-copy sequences in the *X. laevis* genome to sequences in the *X. borealis* genome is difficult to prove because the kinetic retardation observed would require impracticably high *Cots* in order to obtain complete reaction. The interpretation we present here is nonetheless quantitatively consistent with the experimental data.

According to the albumin divergence studies of Wilson and his associates (see Chapter 13), about 25 to 30 MY have elapsed since the divergence of *X. laevis* and *X. borealis*. The gross rate of change in single-copy sequence in the genus *Xenopus* thus seems significantly more rapid than in the primates, which have a genome size only about 20 percent larger than *Xenopus*. Single-copy heteroduplexes formed between the DNAs of primate species whose common ancestor existed 25 to 30 MY ago display only about a 2 to 3°C reduction in melting temperature from that of the homologous duplexes (Kohne, 1970). This is in contrast to the 10 to 13°C depression in heteroduplex melting temperature reported here. It is difficult to account for this large difference in divergence rate between primates and *Xenopus* species on the basis of differences in generation time alone. Whether it is a function of the amphibian DNA replication and repair mechanisms or has other sources cannot at present be decided. In any case it is clear that no safe generalization with regard to the rate of DNA sequence change can be derived from studies on any one taxon.

CONCLUSIONS

Male hybrids of *X. laevis* and *X. borealis* are totally sterile, though some fertile female hybrids can be produced. However, the fertility of these hybrid animals is in a sense an accident of an abnormal meiotic prophase. Though the *X. borealis* and *X. laevis* chromosome in some of the oocytes of these hybrids fail to synapse with each other, the fertile eggs develop from tetraploid primary oocytes. In such oocytes two sets of *X. borealis* and two sets of *X. laevis* chromosomes are later found synapsed each with homologues of its own species. On ovulation and reduction divisions, a diploid egg is formed, which when fertilized by a normal haploid sperm can develop into a mature but triploid adult. This account is based on studies of M. Fischberg and collaborators (personal communication). Its significance here is that by the usual criterion *X. laevis* and *X. borealis* are clearly distinct species. They are closely related morphologically, though key species differences exist. The main point is that in ontogenic terms they are sufficiently compatible to develop into hybrid adults whether fertile or not.

As shown above, a large amount of total DNA sequence divergence has occurred since the evolutionary separation of these two anuran congeners. Brown *et al.* (1972) and Brown and Sugimoto (1973) have shown that the ribosomal and 5S RNA–DNA spacer sequences in the same two species (referred to by them as *X. laevis* and *X. mulleri*) are so different that they fail to display any heterologous cross reaction. One of several possibilities is that the mechanism of such changes in the content of tandem repetitive sequence clusters has been unequal crossover. In this chapter we have seen that most of the single-copy sequence has diverged at a high rate as well. There has been at least 10 to 15 percent nucleotide substitution. Replication error, failure of repair processes, and random insertion of base transitions and transversions rather than unequal crossover are evidently among the responsible mechanisms here. Among the most significant aspects of the findings reviewed here is the conclusion that two genomes capable of cooperating to produce an adult hybrid can tolerate such a large amount of genomic sequence alteration. Only one interpretation seems reasonable, and that is that the specific sequence of much of the single-copy DNA is not functionally required during the life of the animal. This is not to say that this DNA is functionless, only that its specific sequence is not important. Such would be the case, for example, if some of the single-copy sequence served as a form of spacer sequence. Obviously selection will play a direct role in maintaining at least some of the specific sequence of structural gene elements. These are expected to constitute a minor fraction of the single-copy DNA, probably less than 10 percent. Thus we would expect that a small portion of the nonrepetitive sequence must exist which shows far less divergence between the two species, but isolation of the structural gene sequences would be necessary to demonstrate this.

A particularly interesting observation is that the bulk of the repetitive sequences present in each genome is able to cross-react with those of the other genome. It is clear that most of or all the repetitive sequence families are present in both genomes. Two explanations for this conclusion exist: Either these families are required physiologically, or they have been protected by linkage to the single-copy regions in which many of them are embedded so that they could not be lost in the time since divergence. Sharing among related species of the interspersed repetitive sequences has been demonstrated in studies of rodent DNAs as well (Rice, 1972). However, in contrast to the latter studies, the repetitive heteroduplex structures of the two *Xenopus* species display about the same thermal stability as the homologous repetitive duplexes. To some extent this may result from the relatively stringent pairing criterion at which the observations were made. The main fact remains, however, that most of the medium- and higher-stability homologous reaction with repetitive DNA is also found with the heterologous reaction. (For example, see Figure 4E.) If we better understood the function of the interspersed repetitive sequences, it would be easier to interpret the meaning of this result.

The kinetic experiments on the heterologous repetitive DNA reactions shown in Figure 3 lead to the conclusion that the frequency of occurrence of some repetitive sequences has changed since the divergence of the two species. After correcting for the effect of divergence on reaction rate we are still left with the conclusion that a significant fraction of the repetitive sequences appear to have changed in frequency. This fraction may fall in the long repeat class (about 25 percent of the total repetitive sequence), but whether this is the case is unknown. The unequal crossover mechanism provides a means by which the number of copies of given sequences can be rapidly altered, but this explanation pertains only to tandemly repetitive sequence clusters. To explain changes in frequency of occurrence of interspersed repetitive sequences presents a new problem of formidable dimensions.

It now appears that the molecular meaning of DNA sequence organization is necessary for understanding the evolution of the genome. As noted earlier (Britten and Davidson, 1969, 1971), the study of genomic regulation and the study of evolution must be considered two sides of the same coin.

EXPERIMENTAL DETAILS FOR THE FIGURES IN CHAPTER 12

Figure 1. Reactions were carried out in 0.12 to 0.41-*M* phosphate buffer at 60°C and were analyzed by hydroxyapatite chromatography in 0.12-*M* phosphate buffer at 60°C. Incubations to *Cot*s less than 100 to 1,000 were carried out in 0.12-*M* phosphate and higher *Cot* incubations in higher salt concentrations up to 0.41-*M* phosphate buffer. The *Cot* values plotted have been corrected for the accelerating effect of higher salt concentrations (Britten *et al.,* 1974). That is, all *Cot* values presented are calculated as the *Cot* which would have been required to obtain the same amount of reassociation under standard 60°C 0.12-*M* phosphate buffer conditions ("equivalent" *Cot*). The components are as follows: A: 6 percent of the fragments reacting at a rate of 103 M^{-1} sec^{-1}; B: 31 percent of the fragments reacting at 0.72 M^{-1} sec^{-1}; C: 10 percent of the fragments reacting at 0.036 M^{-1} sec^{-1}; D: 54 percent of the fragments reacting at 0.00035 M^{-1} sec^{-1}. The last is the single-copy rate for the *X. laevis* genome. These kinetic components were derived by Davidson *et al.* (1973).

Figure 2. Conditions and data reduction as in Figure 1.

Figure 3. Repetitive *X. laevis* and *X. borealis* ^{3}H DNA were incubated separately with at least 6,600-fold excesses of unlabeled, whole heterologous DNA in 0.12-*M* phosphate buffer at 60°C. The fraction of tracer hybridized was assayed by hydroxyapatite in 0.12-*M* phosphate buffer at 60°C. Self-reaction of the tracers (calculated to be less than 3 percent beyond the very rapid binding material) has not been subtracted from the hybrid data.

A: The components indicated by the dashed curves were obtained by holding the rate of the major fast component at 0.72 M^{-1} sec^{-1}, which is the rate of the major repetitive sequence component in both DNAs (Figure 1). The least-squares rate for the smaller component is 0.004 M^{-1} sec^{-1}. If both rates and quantities are free to change, an equally good fit is obtained with rate constants of 1.02 and 0.003 M^{-1} sec^{-1}.

B: The rate of the faster component was held at 0.72 M^{-1} sec^{-1} in this particular solution, and the best-fit rate for the slower component is then 0.02 M^{-1} sec^{-1}. Equally good fits can be obtained by allowing both rates to change, yielding rates of 1.2 and 0.02 M^{-1} sec^{-1}.

Figure 4. The reaction mixtures were placed over hydroxyapatite in 0.12-*M* phosphate buffer at 60°C and the single-strand DNA fragments washed through the column with 0.12-*M* phosphate buffer. The temperature was then increased in 4 to 6°C increments. The DNA fragments rendered entirely single-stranded at each temperature interval were eluted from the column with 0.12-*M* phosphate buffer.

E: To facilitate the direct comparison, the elution of the *X. laevis–X. borealis* hybrids was slightly adjusted by normalizing the driver elution profiles to that of the *X. laevis* driver DNA and then apportioning the *X. laevis* ^{3}H DNA radioactivity in each temperature increment in the same ratio to its driver DNA as in the original data.

Figure 5. A: Nonrepetitive *X. laevis* ^{3}H DNA was incubated in 0.12 to 0.41-*M* phosphate buffer at 60°C with excess *X. laevis* DNA, and the fraction of nonrepetitive DNA in hybrid was assayed by hydroxyapatite. High *Cot* data are from reactions run in 0.41-*M* phosphate buffer.

B: Nonrepetitive *X. laevis* ^{3}H DNA was incubated with 5,000-fold to 10,000-fold excess of whole *X. borealis* DNA in 0.41-*M* phosphate buffer at 60°C, and the fraction of tracer hybridized was measured by hydroxyapatite. The rate of the small repetitive component is fixed at 0.036 M^{-1} sec^{-1}, which is the rate of the slow repetitive DNA driver component. (See Figure 1.) The second component is a least-squares fit to the data obtained at higher *Cot* values using the expression

$$\frac{G}{G_0} = V^{-.75T} \exp \frac{.25TV^{.56}}{.56}$$

where G and G_0 are the concentrations of single-stranded nonrepetitive tracer at various times and at time zero, respectively; $V = 1 + k\ Cot$, where k is the second-order rate constant of the driver DNA; Co, its concentration; and $t = R/k$, where R is the rate constant for the reaction of the tracer with the driver DNA. In this instance R was fixed at $0.2k$, where k is the homologous driver rate of 0.00036 M^{-1} sec^{-1}. The ratio 0.2 is calculated from the data of Bonner *et al.* (1973) for the incubation conditions and the thermal stability of the hybrids taken as $T_m = 64$°C. The solid curve thus provides the best estimate of the total reaction kinetics, where one takes into account the extent of binding observed and the characteristics of the heterologous duplexes at the highest *Cot*s possible as well as our current understanding of the nature of retarded reaction kinetics.

Figure 6. Nonrepetitive *X. laevis* ^{3}H DNA was incubated separately with excess *X. laevis* DNA or *X. borealis* DNA in 0.41-*M* phosphate buffer at 60°C as described for Figure 5. The reaction mixtures were then applied to hydroxyapatite and thermally eluted as described for Figure 4.

SUGGESTED READINGS

The classic paper on repetitive versus nonrepetitive DNA sequences is R. J. Britten and D. E. Kohne, Repeated sequences in DNA, *Science,* **161,** 529–540, 1968. Two general reviews that should be consulted are R. J. Britten and E. H. Davidson, Repetitive and nonrepetitive DNA sequences and a speculation on the origins of evolutionary novelty, *Quart. Rev. Biol.,* **46,** 111–138, 1971; and E. H. Davidson and R. J. Britten, Organization, transcription, and regulation in the animal genome, *Quart. Rev. Biol.,* **48,** 565–613, 1973.

For an introduction to the techniques of DNA denaturation and reannealing see the paper by Britten and Kohne cited above; also see B. H. Hoyer, B. J. McCarthy, and E. T. Bolton, A molecular approach in the systematics of higher organisms, *Science,* **144,** 959–967, 1964; and C. D. Laird and B. J. McCarthy, Magnitude of interspecific nucleotide sequence variability in *Drosophila, Genetics,* **60,** 303–322, 1968. These two articles may also serve as introductions to the use of DNA hybridization in phylogenetic studies. The application of this method to the study of primate evolution is the subject of D. E. Kohne, J. A. Chiscon, and B. H. Hoyer, Evolution of primate DNA sequences, *J. Hum. Evol.,* **1,** 627–644, 1972.

CHAPTER 13

GENE REGULATION IN EVOLUTION

A. C. WILSON

One much-argued problem of evolutionary genetics during the last decade was the possibility that sequence changes in proteins result mainly from adaptively neutral mutations. This problem, however, is central to evolutionary genetics only if adaptive evolution is based predominantly on changes in genes coding for proteins, i.e., structural genes. This assumption remains untested. Indeed, it is possible that regulatory mutations play the major part in adaptive evolution. This chapter evaluates the relative importance of these two types of genetic change. Although definitive conclusions are not possible at present, it seems likely that evolution at the organismal level depends predominantly on regulatory mutations. Structural gene mutations may have a secondary role in organismal evolution.

Regulatory mutations influence the expression of genes coding for certain proteins without necessarily affecting the structure of those proteins. Two types of regulatory mutations may be considered: First, mutations can occur in regulatory genes. Such mutations can affect dramatically the production though not the amino acid sequences of specific groups of enzymes. Second, the order of genes on chromosomes may change because of inversion, translocation, duplication, or deletion of genes as well as fusion or fission of chromosomes. (See Chapter 1.) These events can bring genes into new relationships with one another, occasionally resulting in altered patterns of gene expression (Bahn, 1971; Wallace and Kass, 1974).

REGULATORY CHANGES IN BACTERIAL ADAPTATION

Direct evidence for the evolutionary importance of regulatory mutations comes from experiments with bacterial populations. Bacteria are in several ways ideal organisms with which to study the mechanism of evolution. First, bacterial populations adapt rapidly in the laboratory to new situations. This is because one can work with large populations having short generation times. Second, bacteria are relatively simple genetically, having only a few thousand genes, in contrast to vertebrates which may have more than a million genes per cell. Third, the biochemistry and genetics of some bacteria are so well known that one can hope to gain a precise molecular understanding of how they adapt to a well-defined change in environment. Much progress toward such an understanding has come from studies in which the environmental change is the appearance of a novel chemical.

When a bacterial population encounters a novel carbon compound, rare individuals may by chance carry a mutation that permits metabolic utilization of the compound. The mutants have a selective advantage if no other carbon source is available. Laboratory studies reveal that the primary event permitting such adaptation to a new resource is often a regulatory mutation. The regulatory mutant produces a high concentration of an enzyme that has weak activity on the new compound because of chance chemical resemblance between the latter and the normal substrate. By virtue of having perhaps 100 times more of this enzyme than the wild-type bacteria do, the mutant can metabolize the new compound at a biologically significant rate Lerner *et al.*, 1964; Wilson, 1975, and references therein).

A second stage of genetic adaptation to a novel carbon source can sometimes result from mutations affecting the amino acid sequence of a derepressed enzyme. The mutations, occurring in the gene coding for the enzyme, may alter the active site so that its reactivity with the novel analogue increases. The altered enzyme can therefore catalyze more rapidly the metabolism of the new compound. In conclusion, there is little doubt that regulatory mutations are of primary importance in permitting bacteria to utilize new resources.

RATES OF EVOLUTION

Many biologists have observed a correlation between genic similarity estimated by protein comparisons, and organismal similarity measured in terms of taxonomic distance. Selander and Johnson (1973) and Avise (1974) have summarized electrophoretic evidence for such a correlation. (See Chapter 7.) However, this correlation could result simply from the fact that both structural genes and anatomy usually evolve at fairly steady rates. If genic change and organismal change are each correlated with time, they will seem to be correlated with each other. To find out whether organismal change is dependent on structural gene mutations, one must compare the rates of structural gene change in taxonomic groups which have experienced contrasting rates of organismal change.

The vertebrates are well suited for such a study. Some vertebrate lineages have experienced faster rates of phenotypic evolution than others. Placental mammals, for instance, have experienced rapid organismal evolution compared with lower vertebrates, of which frogs are a typical example. Although there are thousands of frog species living today (Gorham, 1974), they are so uniform phenotypically that zoologists put them all in a single order (Anura), whereas placental mammals are divided into at least 16 orders (Table I). The anatomical diversity represented by bats, whales, cats, and people is unparalleled among frogs, yet frogs are a much older group than placental mammals. By way of illustration, the frog genus *Xenopus* was in existence before the radiation of the placental mammals began (Estes, 1975). Organismal evolution has been slow in frogs relative to mammals. Any type of genetic change that has evolved rapidly in mammals but slowly in frogs might be at the basis of organismal evolution.

Protein Evolution in Frogs and Mammals

As shown in Chapters 9 and 10, rates of protein evolution are measured by comparing the amino acid sequences of homologous proteins from species of known divergence time. Quantitative immunological comparisons provide a simple and economical way of estimating the approximate extent of sequence difference among proteins. A mod-

TABLE I. Rates of evolution in frogs and placental mammals

Property	**Frogs**	**Placental mammals**
Number of living species	3,050*	4,600
Number of orders	1	16–20
Age of the group (MY)	150	75
Rate of organismal evolution	Slow	Fast
Rate of albumin evolution	Standard	Standard
Rate of loss of hybridization potential	Slow	Fast
Rate of change in chromosome number	Slow	Fast
Rate of change in number of chromosomal arms	Slow	Fast

***This figure for the number of frog species known in 1970 is taken from Gorham (1974) and may underestimate rather grossly the actual number of living species because so many frog species are being discovered every year.**

erately strong correlation ($r = 0.9$) exists between degree of sequence difference and degree of antigenic difference measured by the micro-complement fixation technique (Figure 1, Chapter 10; see also Champion *et al.*, 1974, 1975). This technique was used by Maxson, Sarich and Wallace in my laboratory to compare the albumins of hundreds of frog and mammal species. Albumin is an ideal protein for evolutionary studies because it evolves quite rapidly and consists of one polypeptide chain that is 580 amino acids long; thus the albumin gene is equivalent in size to the combined genes for cytochrome *c*, myoglobin, the hemoglobin chains (alpha and beta), and the fibrinopeptides A and B. Albumin seems to have evolved no faster in mammals than in frogs (Maxson *et al.*, 1975; and references therein). Species which are similar enough in anatomy and way of life to be included within a single genus of frogs (e.g., *Rana*) can differ as much in their albumins as does a bat from a whale.

Other proteins have been studied less extensively, although the results are consistent with those obtained from albumin. The beta chains of hemoglobin from two species of *Rana* differ in their amino acid sequences by at least 29 substitutions, which is a greater difference than that usually found between any two orders of placental mammals (Baldwin and Riggs, 1974). Additional hemoglobin evidence (Maxson and Wilson, 1975) as well as electrophoretic evidence obtained with numerous enzymes (Case *et al.*, 1975) agrees with the hypothesis that anatomically similar frogs can differ greatly at the protein sequence level.

DNA Evolution in Frogs and Mammals

DNA annealing studies with nonrepeated sequences have been conducted with a variety of mammals (Kohne *et al.*, 1972; Hoyer *et al.*, 1972; Beneveniste and Todaro, 1974) and two sibling species of frog belonging to the archaic genus *Xenopus* (Chapter 12). The latter two species are so similar at the organismal level that they were both included in *X. laevis* (as the subspecies *X. l. laevis* and *X. l. borealis*) until very recently. The *Xenopus* difference ($\Delta T_m = 12°C$) is larger than that found between the DNA of humans and New World monkeys ($\Delta T_m = 10°C$). These limited studies give one no reason to think that sequence evolution at the DNA level has been slower in frogs than in mammals.

Regulatory Evolution in Frogs and Mammals

Two lines of evidence are suggestive of sluggish regulatory evolution in frogs compared with mammals. To appreciate the first line of evidence, one needs to consider interspecific hybridization. Although there are many natural barriers to fertilization of an egg by sperm of another

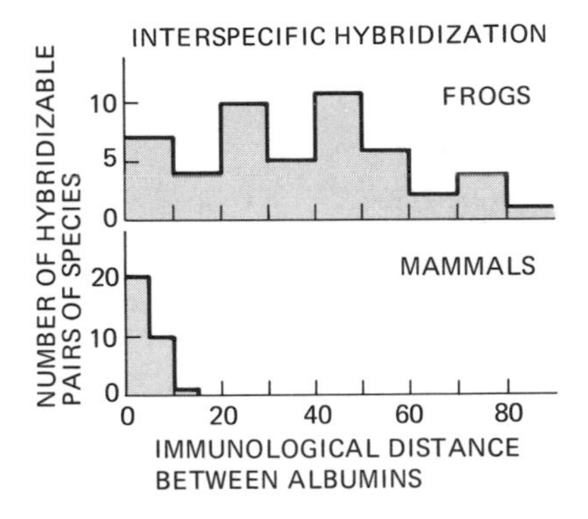

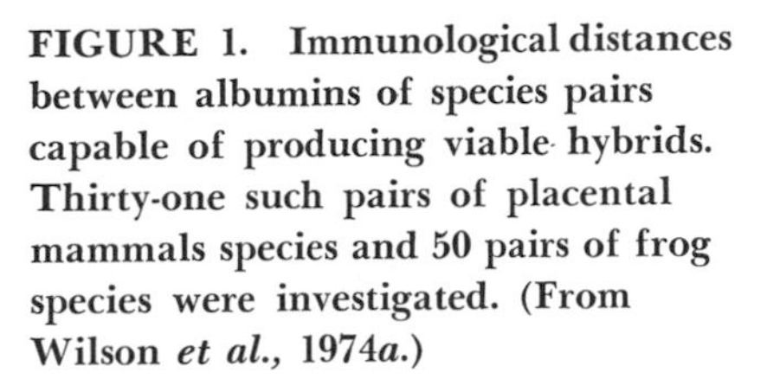
FIGURE 1. Immunological distances between albumins of species pairs capable of producing viable hybrids. Thirty-one such pairs of placental mammals species and 50 pairs of frog species were investigated. (From Wilson *et al.*, 1974*a*.)

species, these are usually not absolute. Supposing an interspecific zygote is formed, one can ask what chance it has of developing into a healthy adult. Embryonic development involves an orderly program of expression of many genes that were inactive in the zygote. If the two genomes in an interspecific zygote are similarly programmed, so that a given block of genes will be turned on at the same time in one genome as in the other, orderly development of a hybrid organism can be expected. However, should the patterns of gene activation differ, the probability of an interspecific zygote developing successfully would be low. In accordance with this view, organismal hybrids derived from extremely different parental species often show signs of breakdowns in gene regulation (Ohno, 1969; Whitt *et al.*, 1973).

If regulatory evolution has proceeded slowly in frogs relative to mammals, one would expect that frog species should retain the ability to hybridize with one another much longer than mammals do. Because the rate of albumin evolution in frogs has been equal to that in mammals, one would expect to find small albumin distances among mammals capable of hybridizing, whereas among hybridizable frogs one should encounter large immunological distances. The albumin results are in accordance with this expectation (Figure 1).

Chromosomal studies provide a second line of evidence consistent with slower regulatory evolution in frogs than in mammals. Placental mammals have experienced far more rapid karyotypic change than have frogs. This is evident from albumin studies on hundreds of species of known karyotype (Wilson *et al.*, 1974*b*). The albumin immunological distance at which there is a 50 percent chance that two species will differ in chromosome number is about 6 units for mammals and about 120 units for frogs (Figure 2). Changes in chromosomal number are brought about by fusion or fission events, while inversions may result in changes in the number of chromosome arms (Chapter 1). The rates at which arm

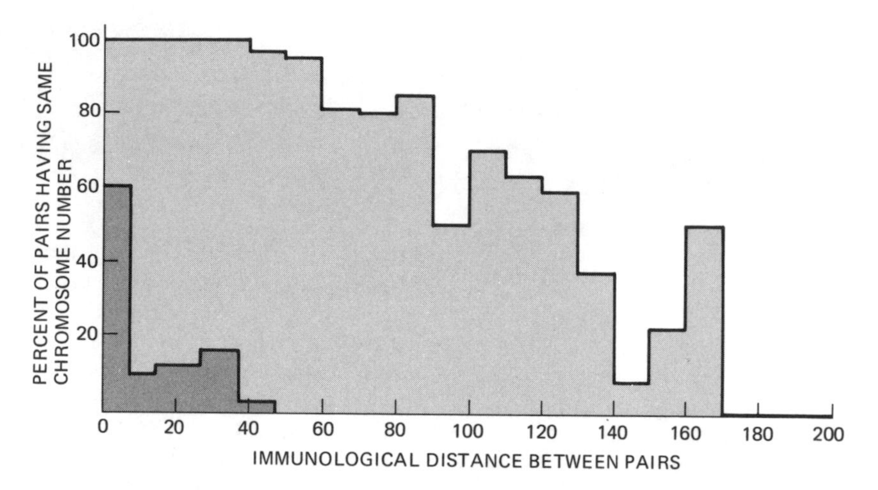

FIGURE 2. Proportion of species pairs having identical chromosome number as a function of the immunological distance between the albumins of the pairs. The light gray histogram summarizes the results for 373 different pairs of frog species. The darker histogram summarizes the results for 318 different pairs of placental mammal species. (From Wilson *et al.*, 1974*b*.)

number has changed turn out to be very similar to the rates of change in chromosome number (Wilson *et al.*, 1974*b*). It is inferred that these two distinct types of gene rearrangement have each been evolving an order of magnitude faster in placental mammals than in frogs.

The studies on rates of evolution in frogs and mammals suggest that sequence changes in structural genes occur to a large extent independently of organismal change whereas regulatory changes as manifested by studies of karyotype and hybrid viability evolve in parallel with organismal change.

Evolution in Humans and Chimpanzees

The comparison of human and chimpanzee macromolecules supports the conclusion inferred in the previous paragraph. Because of major differences at the organismal level, these two species are classified in different taxonomic families, yet at the macromolecular sequence level humans and chimpanzees are extraordinarily similar. Species within a genus of mice, frogs, or flies can differ more from each other than does a human from a chimpanzee, as is evident from the DNA studies summarized in Table II. A similar result is obtained from extensive protein comparisons, which are illustrated in Figure 3. It seems probable that at the structural gene level chimpanzees and humans are as similar as a pair of sibling species (King and Wilson, 1975).

An evolutionary perspective further illustrates the contrast between

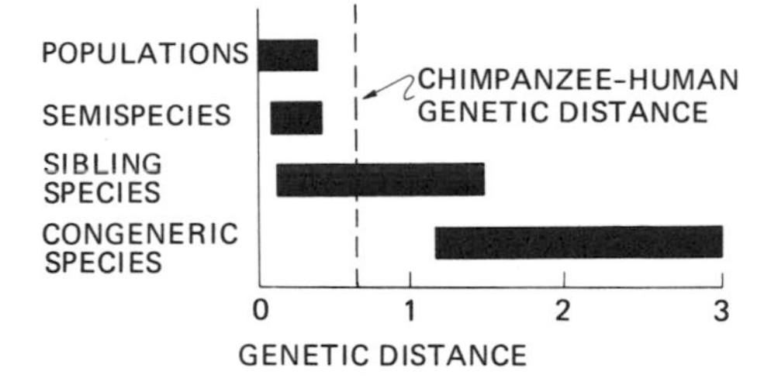

FIGURE 3. The genetic distance between humans and chimpanzees compared with that between other organisms. These estimates of genetic distance are based on electrophoretic comparison of many proteins. One unit of genetic distance means that there is an average of one electrophoretically detectable substitution per polypeptide compared. The bars indicate the range of genetic distances found among a wide variety of organisms, including fruit flies, horseshoe crabs, salamanders, lizards, fish, bats, and rodents.

the results of the molecular and organismal approaches. Since the common ancestor of the two species lived, the chimpanzee lineage has evolved slowly relative to the human lineage, in terms of anatomy and adaptive strategy. This concept is illustrated in Figure 4. However, phylogenetic analysis of the sequence comparisons among chimpanzees, humans, and other primates indicates that the two lineages have undergone approximately equal amounts of sequence change in their macromolecules. Hence, the major adaptive change which took place in the human lineage was probably not accompanied by accelerated DNA or protein sequence evolution.

What then is the genetic basis for the evolution of humans from apes?

TABLE II. DNA comparisons made by measuring the thermal stability of hybrid double strands

Species compared	Sequence difference in DNA* (%)	Level of taxonomic difference
Hominoids		
Human with chimpanzee	1.1	Family
Mice		
Mus musculus with *M. cervicolor*	5	Species
Frogs		
Xenopus laevis with *X. borealis*	12	Species
Flies		
Simulium pictipes with *S. venustum*	11	Species
Drosophila melanogaster with *D. salmon*	19	Species

*Estimated from studies with nonrepetitive DNA and the assumption that a 1 percent difference in base sequence lowers thermal stability by 1°C. From Wilson (1975).

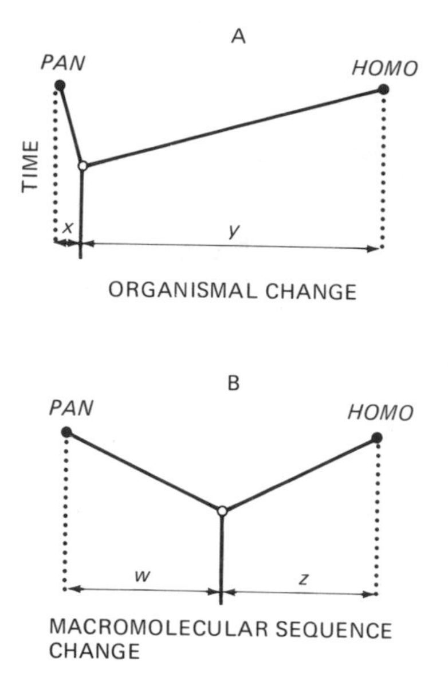

FIGURE 4. The contrast between biological evolution and molecular evolution since the divergence of the human and chimpanzee lineages. A: More biological change has taken place in the human lineage (*y*) than in the chimpanzee lineage (*x*). B: Protein and nucleic acid evidence indicate that as much change has occurred in chimpanzee genes (*w*) as in human genes (*z*). (From King and Wilson, 1975.)

The organismal differences between apes and humans might result chiefly from genetic changes in a few regulatory systems. The idea that regulatory mutations have had a major role in human evolution is consistent with the old observation that adult humans resemble fetal apes in some respects. The retention of fetal patterns of gene expression during childhood may have played a part in evolving such human features as our large brain, our small jaws and canine teeth, our naked skin, and our upright posture (Gould, 1975).

Gene rearrangement could be an important source for such changes in gene regulation during human evolution. Although humans and chimpanzees do not differ greatly in chromosome number, having 46 and 48 chromosomes, respectively, the arrangement of genes differs in the two species. Only a small proportion of the chromosomes has identical banding patterns in these two hominoids. Banding studies indicate that at least 10 large inversions and translocations and one chromosomal fusion have occurred since the two lineages separated. Further evidence showing that humans and chimpanzees differ considerably in chromosomal organization emerges from studies in which a purified nucleic acid fraction was annealed to chromosomes *in situ* (King and Wilson, 1975). The chromosomal sites at which annealing occurred were distributed quite differently in the two hominoids, indicating different gene arrangements on the chromosomes. It is therefore proposed, as a working hypothesis, that chimpanzee and human genes are remarkably similar simply because the species separated rather recently (Sarich and Wilson, 1973; Goodman, 1974) whereas the large organismal differences are due to rapid regulatory evolution in the human lineage.

Evolution in Other Organisms

Many other examples of contrasts between genic similarity and organismal similarity are known in a wide variety of taxonomic groups, ranging from bacteria (Stanier *et al.*, 1970) and ciliates (Borden *et al.*, 1973) to snails (Gould *et al.*, 1974), fish (Turner, 1974), frogs (Maxson and Wilson, 1974, 1975), reptiles (Mao and Dessauer, 1971), and birds (Nolan *et al.*, 1975). It may therefore be a general rule that organismal evolution and structural gene evolution go on at virtually independent rates. This is also supported by the intriguing evidence that speciation can occur without alteration of structural genes in both insects (Bush, 1975) and flowering plants (Chapter 8).

There is at present little published evidence for a correlation between organismal evolution and evolution at the level of chromosome organization or hybrid viability for groups other than frogs and placental mammals. Some evidence is available for birds. Relative to placental mammals, this group has experienced slow anatomical evolution in the past 30 MY. During this period, birds have experienced slow loss of hybridization potential and slow chromosomal evolution (Prager and Wilson, 1975) compared with placental mammals.

CONCLUSIONS

The following conclusions are drawn tentatively from the data reviewed above concerning the mechanism of bacterial adaptation and rates of evolution. Evolution at the organismal level may depend primarily on regulatory mutations, which alter patterns of gene expression. Mutations affecting the arrangement of genes on chromosomes may be a common source of these altered patterns of gene expression. By contrast, sequence change in structural genes may be of secondary importance; this process is not tightly geared to organismal evolution and tends to go on relentlessly at about the same rate in all organisms.

To test these tentative conclusions, it is important to study in greater depth the genetic and biochemical mechanisms underlying bacterial adaptation. More thorough studies of rates of evolutionary change in anatomy, genes, chromosomes, and hybrid viability are also relevant. In order to examine quantitatively the relationship between molecular evolution and organismal evolution, it is important to use numerical methods for estimating the degree of organismal resemblance. Until we know how much more rapidly placental mammals have evolved than frogs have at the organismal level, we cannot know how strong the correlation is between organismal change and change in chromosomes or in hybridization potential. It is also important for evolutionary geneticists to become familiar with current research on the organization of eukaryotic genomes and on the molecular biology of gene expression, especially during embryogenesis. Interaction with these areas of research would stimulate the development of better methods for estimating degree of difference between patterns of gene expression.

Comparative studies of amino acid and nucleotide sequences will also remain important. Sequences evolve at fairly steady rates that are virtually independent of rates of organismal evolution (Maxson *et al.,* 1975, also, Chapter 10). Proteins and nucleic acids may become valuable tools for elucidating the order of branching of lineages and for estimating the approximate times of divergence of lineages (Sarich, 1973; Maxson and Wilson, 1975).

SUGGESTED READINGS

The classical model of gene regulation in bacteria is presented in F. Jacob and J. Monod, Genetic regulatory mechanisms in the synthesis of proteins, *J. Mol. Biol.,* **3,** 318–356, 1971.

The following are examples of authors that considered the possible evolutionary role of regulatory mutations: B. Wallace, Genetic diversity, genetic uniformity, and heterosis, *Canad. J. Genet. Cytol.,* **5,** 239–253, 1963; E. Zuckerkandl, Perspectives in molecular anthropology, in S. L. Washburn (ed.) *Classification and Human Evolution,* Wenner-Gren Foundation, New York, 1963; and G. L. Stebbins, *The Basis of Progressive Evolution,* University of North Carolina Press, Chapel Hill, 1969.

Direct evidence for the importance of regulatory mutations in adaptive evolution was provided in a classic paper by S. A. Lerner, T. T. Wu and E. C. C. Lin, Evolution of a catabolic pathway in bacteria, *Science,* **146,** 1313–1315, 1964.

An additional stimulus to consider the role of regulatory mutations in evolution came from the models of gene regulation in eukaryotes proposed by R. J. Britten and E. H. Davidson, Gene regulation for higher cells: a theory, *Science,* **165,** 349–357, 1969; S. Ohno, An argument for the genetic simplicity of man and other mammals, *J. Human Evol.* **1,** 651–662, 1972; and B. Wallace and T. Kass, On the structure of gene control regions, *Genetics,* **77,** 541–558, 1974.

Evidence regarding the relative importance of regulatory mutations and structural gene mutations for evolution at the organismal level has been analysed by A. C. Wilson and his colleagues in several publications, such as M.–C. King and A. C. Wilson, Evolution at two levels in humans and chimpanzees, *Science,* **188,** 107–116, 1975; and A. C. Wilson, Evolutionary importance of gene regulation, *Stadler Genet. Sympos.,* **7,** 117–134, 1975.

ACKNOWLEDGMENTS

This book is based on a symposium, "Molecular Study of Biological Evolution," held at the University of California, Davis, June 17 and 18, 1975. The symposium was sponsored by the Society for the Study of Evolution and the American Society of Naturalists and was made possible by a generous grant from the Ernest Henderson Foundation of Wellesley, Massachusetts.

Ayala: Molecular Genetics and Evolution

The author is indebted to Dr. Gordon J. Edlin for critical reading of the manuscript. The author's research is supported by grants from the National Science Foundation and the Energy Research and Development Administration.

Selander: Genic Variation in Natural Populations

M. Sepkoski assisted in preparation of the manuscript. The author's research is supported by grants from the National Science Foundation and the National Institutes of Health.

Soulé: Allozyme Variation: Its Determinants in Space and Time

The author is indebted to Michael Gilpin and Christopher Wills, who read and criticized the manuscript; to Jack L. King and Timothy Prout, who were very patient in explaining population genetics and suggesting useful literature; and to George Gorman and Suh Yung Yang for contributing advice and data.

Valentine: Genetic Strategies of Adaptation

The author is indebted to F. J. Ayala and C. A. Campbell for critical reading of the manuscript, and to D. Hedgecock and W. W. Anderson for providing unpublished estimates of genetic variability in some marine invertebrates.

Goodman: Protein Sequences in Phylogeny

The author's research is supported by a grant from the National Science Foundation.

Fitch: Molecular Evolutionary Clocks

The author gratefully acknowledges the assistance of Dr. R. L. Niece. The author's research is supported by grants from the National Science Foundation and the National Institutes of Health.

Galau, Chamberlin, Hough, Britten, and Davidson: Evolution of Repetitive and Nonrepetitive DNA

The authors are indebted to Dr. D. Brown for his generous gift of *Xenopus* DNAs. The authors' research is supported by grants from the National Institutes of Health and the National Science Foundation.

Wilson: Gene Regulation in Evolution

A largely similar presentation of the ideas advanced in this chapter may be found in volume 7 (1975) of the *Stadler Genetics Symposia*. The author is indebted to S. M. Beverly, S. S. Carlson, V. M. Sarich, and T. J. White for helpful discussions. This chapter is dedicated to Professor E. C. C. Lin and his students, who published in 1964 a classic article demonstrating the evolutionary importance of gene regulation.

BIBLIOGRAPHY

Allard, R. W., 1965, Genetic systems associated with colonizing ability in predominantly self-pollinated species, in *The Genetics of Colonizing Species,* H. G. Baker and G. L. Stebbins (eds.), Academic Press, New York, pp. 49–75.

Allard, R. W., S. K. Jain, and P. L. Workman, 1968, The genetics of inbreeding populations, *Adv. Genet.,* **14,** 55–131.

Allard, R. W., and A. L. Kahler, 1971, Allozyme polymorphisms in plant populations, *Stadler Symp.* (University of Missouri), **3,** 9–24.

Allard, R., and A. Kahler, 1973, Multilocus genetic organization and morphogenesis, *Brookhaven Symp. Biol.,* **25,** 329–343.

Allard, R. W., A. L. Kahler, and M. T. Clegg, 1975, Isozymes in plant population genetics, in *Isozymes. IV. Genetics and Evolution,* C. L. Markert (ed.), Academic Press, New York, pp. 261–272.

Angerer, R. C., E. H. Davidson, and R. J. Britten, 1975, DNA sequence organization in the mollusc *Aplysia californica, Cell,* **6,** 29–39.

Antonovics, J., 1968, Evolution in closely adjacent plant populations. V. Evolution of self-fertility, *Heredity,* **23,** 219–238.

Atkinson, D., 1971, Adenine nucleotides as stoichiometric coupling agents in metabolism and as regulatory modifiers. The adenylate energy charge, in *Metabolic Pathways,* vol. V, H. Vogel (ed.), Academic Press, New York, pp. 1–21.

Avery, O. T., C. M. MacLeod, and M. McCarty, 1944, Studies on the chemical nature of the substance inducing transformation of pneumococcal types. Induction of transformation by a deoxyribonucleic acid fraction isolated from pneumococcus Type III, *J. Exp. Med.,* **79,** 137–158.

Avise, J. C., 1974, Systematic value of electrophoretic data, *System. Zool.,* **23,** 465–481.

Avise, J. C., and F. J. Ayala, 1975*a,* Genetic differentiation in speciose versus depauperate phylads: evidence from the California minnows, *Evolution,* in press.

Avise, J. C., and F. J. Ayala, 1975*b,* Genetic change and rates of cladogenesis, *Genetics,* **81,** 757–773.

Avise, J. C., and G. G. Kitto, 1973, Phosphoglucose isomerase gene duplication in the bony fishes: an evolutionary history, *Biochem. Genet.,* **8,** 113–132.

Avise, J. C., and R. K. Selander, 1972, Evolutionary genetics of cave-dwelling fishes of the genus *Astyanax, Evolution,* **26,** 1–19.

Avise, J. C., and M. H. Smith, 1974*a,* Biochemical genetics of sunfish. I. Geographic variation and subspecific intergradation in the bluegill, *Lepomis macrochirus, Evolution,* **28,** 42–56.

Avise, J. C., and M. H. Smith, 1974*b*, Biochemical genetics of sunfish. II. Genic similarity between hybridizing species, *Amer. Natur.,* **108,** 458–472.

Avise, J. C., M. H. Smith, and R. K. Selander, 1974*a*, Biochemical polymorphism and systematics in the genus *Peromyscus.* VI. The *boylii* species group, *J. Mammal.,* **55,** 751–763.

Avise, J. C., M. H. Smith, R. K. Selander, T. E. Lawlor, and P. R. Ramsey, 1974*b*, Biochemical polymorphism and systematics in the genus *Peromyscus.* V. Insular and mainland species of the subgenus *Haplomylomys, System. Zool.,* **23,** 226–238.

Ayala, F. J., 1965, Evolution of fitness in experimental populations of *Drosophila serrata, Science,* **150,** 903–905.

Ayala, F. J., 1966, Evolution of fitness. I. Improvement in the productivity and size of irradiated populations of *Drosophila serrata* and *Drosophila birchii, Genetics,* **53,** 883–895.

Ayala, F. J., 1968, Genotype, environment, and populations numbers, *Science,* **162,** 1453–1459.

Ayala, F. J., 1969, Evolution of fitness. V. Rate of evolution in irradiated populations of *Drosophila, Proc. Natl. Acad. Sci. U.S.A.,* **63** (3), 790–793.

Ayala, F. J., 1972, Darwinian *versus* non-Darwinian evolution in natural populations of *Drosophila, Proc. Sixth Berkeley Symp. Math. Stat. Prob.,* vol. V, 211–236.

Ayala, F. J., 1975, Genetic differentiation during the speciation process, in *Evolutionary Biology,* vol. 8, T. Dobzhansky, M. K. Hecht, and W. C. Steere (eds.), Plenum Press, New York, pp. 1–78.

Ayala, F. J., and T. Dobzhansky, 1974, A new subspecies of *Drosophila pseudoobscura* (Diptera: Drosophilidae), *Pan. Pac. Entomol.,* **50,** 211–219.

Ayala, F. J., and M. E. Gilpin, 1974, Gene frequency comparisons between taxa: support for the natural selection of protein polymorphisms, *Proc. Natl. Acad. Sci. U.S.A.,* **71,** 4847–4849.

Ayala, F. J., D. Hedgecock, G. S. Zumwalt, and J. W. Valentine, 1973, Genetic variation in *Tridacna maxima,* an ecological analog of some unsuccessful evolutionary lineages, *Evolution,* **27,** 177–191.

Ayala, F. J., and J. R. Powell, 1972, Enzyme variability in the *Drosophila willistoni* group. VI. Levels of polymorphism and the physiological function of enzymes, *Biochem. Genet.,* **7,** 331–345.

Ayala, F. J., J. R. Powell, and T. Dobzhansky, 1971, Polymorphisms in continental and island populations of *Drosophila willistoni, Proc. Natl. Acad. Sci. U.S.A.,* **68,** 2480–2483.

Ayala, F. J., and M. L. Tracey, 1974, Genetic differentiation within and between species of the *Drosophila willistoni* group, *Proc. Natl. Acad. Sci. U.S.A.,* **71,** 999–1002.

Ayala, F. J., M. L. Tracey, L. G. Barr, and J. G. Ehrenfeld, 1974*a*, Genetic and reproductive differentiation of the subspecies *Drosophila equinoxialis caribbensis, Evolution,* **28,** 24–41.

Ayala, F. J., M. L. Tracey, L. G. Barr, J. F. McDonald, and S. Pérez-Salas, 1974*b*, Genetic variation in natural populations of five *Drosophila* species and the hypothesis of the selective neutrality of protein polymorphisms, *Genetics,* **77,** 343–384.

Ayala, F. J., M. L. Tracey, D. Hedgecock, and R. C. Richmond, 1974*c*, Genetic differentiation during the speciation process in *Drosophila, Evolution,* **28,** 576–592.

Ayala, F. J., and J. W. Valentine, 1974, Genetic variability in the cosmopolitan deep-water ophiuran *Ophiomusium lymani, Mar. Biol.* **27,** 51–57.

Ayala, F. J., J. W. Valentine, T. E. DeLaca, and G. S. Zumwalt, 1975*a*, Genetic variability of the Antarctic brachiopod *Liothyrella notorcadensis* and its bearing on mass extinction hypotheses, *J. Paleontol.,* **49,** 1–9.

Ayala, F. J., J. W. Valentine, D. Hedgecock, and L. G. Barr, 1975*b*, Deep-sea asteroids: high genetic variability in a stable environment, *Evolution,* **29,** 203–212.

Ayala, F. J., J. W. Valentine, and G. S. Zumwalt, 1975*c*, An electrophoretic study of the Antarctic zooplankton *Euphausia superba, Limnology and Oceanography,* **20,** 635–640.

Baba, M. L., M. Goodman, H. Dene, and G. W. Moore, 1975, Origins of the Ceboidea viewed from an immunological perspective, *J. Hum. Evol.,* **4,** 89–102.

Bachmann, K., 1972, Genome size in mammals, *Chromosoma,* **37,** 85–93.

Bachmann, K., O. B. Goin, and C. J. Goin, 1974, Nuclear DNA amounts in vertebrates, in *Evolution of Genetic Systems,* H. H. Smith *et al.* (eds.), *Brookhaven Symp. Biol.,* **23,** 419–450.

Bachmann, K., B. A. Harrington, and J. P. Craig, 1972, Genome size in birds, *Chromosoma,* **37,** 405–416.

Bachmann, K., and E. L. Rheinsmith, 1973, Nuclear DNA amounts in pacific *crustacea, Chromosoma,* **43,** 225–236.

Baetcke, K. P., A. H. Sparrow, C. H. Nauman, and S. S. Schwemmer, 1967, The relationship of DNA content to nuclear and chromosome volume and to radiosensitivity (LD_{50}), *Proc. Natl. Acad. Sci. U.S.A.,* **58,** 533–540.

Bahn, E., 1971, Position-effect variegation for an isoamylase in *Drosophila melanogaster, Hereditas,* **67,** 79–82.

Baldwin, T. O., and A. Riggs, 1974, The hemoglobins of the bullfrog, *Rana catesbeiana.* Partial amino acid sequence of the β chain of the major adult component, *J. Biol. Chem.,* **249,** 6110–6118.

Barber, H. N., 1970, Hybridization and the evolution of plants, *Taxon,* **19,** 154–160.

Baur, E., 1925, Die Bedeutung der Mutationen für das Evolutionsproblem, *Zeit. ind. Abst. Verebungs.,* **37,** 107–115.

Baur, E., 1932, Artumgrenzung und Artbildung in der Gattung Antirrhinum, *Zeit. ind. Abst. Verebungs.,* **63,** 256–302.

Beard, J. M., 1975, The haemoglobins of *Tarsius bancus,* paper prepared for *Burg Wartenstein Symposium No. 65, Progress in Molecular Anthropology.*

Beardmore, J., and L. Levine, 1963, Fitness and environmental variation. I. A study of some polymorphic populations of *Drosophila pseudoobscura, Evolution,* **17,** 121–129.

Beckman, G., and L. Beckman, 1975, Genetics of human superoxide dismutase isozymes, in *Isozymes. IV. Genetics and Evolution,* C. L. Markert (ed.), Academic Press, New York, pp. 781–795.

Benesch, R., and R. E. Benesch, 1974, Homos and heteros among the hemos, *Science,* **185,** 905–908.

Beneveniste, R. E., and G. J. Todaro, 1974, Evolution of type C viral genes. I. Nucleic acid from baboon type C virus as a measure of divergence among primate species, *Proc. Natl. Acad. Sci. U.S.A.,* **71,** 4513–4518.

Berger, D., 1974, Esterases of *Drosophila.* II. Biochemical studies of esterase-5 in *D. pseudoobscura, Genetics,* **78,** 1157–1172.

Berger, E. M., and L. Weber, 1974, The ribosomes of *Drosophila.* II. Studies on intraspecific variation, *Genetics,* **78,** 1173–1183.

Bernstein, S. C., L. H. Throckmorton, and J. L. Hubby, 1973, Still more genetic variability in natural populations, *Proc. Natl. Acad. Sci. U.S.A.,* **70,** 3928–3931.

Bier, V. K., and W. Müller, 1969, DNA-Messungen bei Insekten und eine Hypothese über retardierte Evolution und besonderen DNS-Reichtum im Tierreich, *Biol. Zentralbl.,* **88,** 425–449.

Birnstiel, M., J. Telford, E. Weinberg, and D. Stafford, 1974, Isolation and some properties of the genes coding for histone proteins, *Proc. Natl. Acad. Sci. U.S.A.,* **71,** 2900–2904.

Bonnell, M. L., and R. K. Selander, 1974, Elephant seals: genetic variation and near extinction, *Science,* **184,** 908–909.

Bonner, J., W. T. Garrard, J. Gottesfeld, D. S. Holmes, J. S. Sevall, and M. Wilkes, 1974, Functional organization of the mammalian genome, *Cold Spring Harbor Symp. Quant. Biol.,* **38,** 303–310.

Bonner, T. I., D. J. Brenner, B. R. Neufeld, and R. J. Britten, 1973, Reduction in the rate of DNA reassociation by sequence divergence, *J. Mol. Biol.,* **81,** 123–135.

Borden, D., G. S. Whitt, and D. L. Nanney, 1973, Electrophoretic characterization of classical *Tetrahymena pyriformis* strains, *J. Protozool.,* **20,** 693–700.

Brehm, B., and M. Ownbey, 1965, Variation in chromatographic patterns in the *Tragopogon dubius-pratensis-porrifolius* complex (Compositae), *Amer. J. Bot.,* **52,** 811–818.

Bretsky, P. W., and D. M. Lorenz, 1969, Adaptive response to environmental stability: a unifying concept in paleoecology, *Proc. North Amer. Paleontol. Conv.,* pt. E, 522–550.

Briscoe, D. A., A. Robertson, and J. Malpica, 1975, Dominance at *ADH* locus in response of adult *Drosophila melanogaster* to environmental alcohol, *Nature,* **255,** 148–149.

Britten, R. J., and E. H. Davidson, 1969, Gene regulation for higher cells: a theory, *Science,* **165,** 349–357.

Britten, R. J., and E. H. Davidson, 1971, Repetitive and non-repetitive DNA sequences and a speculation on the origins of evolutionary novelty, *Quart. Rev. Biol.,* **46,** 111–138.

Britten, R. J., and E. H. Davidson, 1975, DNA sequence arrangement and preliminary evidence on its evolution, *Fed. Proc.,* in press.

Britten, R. J., D. E. Graham, and B. R. Neufeld, 1974, Analysis of repeating DNA sequences by reassociation, in *Methods in Enzymology,* vol. 29, pt. E, L. Grossman and K. Moldave (eds.), Academic Press, New York, pp. 363–418.

Britten, R. J., and D. E. Kohne, 1968, Repeated sequences in DNA, *Science,* **161,** 529–540.

Brown, D. D., and J. B. Gurdon, 1964, Absence of ribosomal RNA synthesis in the anucleolate mutant of *Xenopus laevis, Proc. Natl. Acad. Sci. U.S.A.,* **51,** 139–146.

Brown, D. D., and K. Sugimoto, 1973, The 5S DNAs of *Xenopus laevis* and *Xenopus mulleri:* the evolution of a gene family, *J. Mol. Biol.,* **78,** 397–415.

Brown, D. D., and K. Sugimoto, 1974, The structure and evolution of ribosomal and 5S DNAs in *Xenopus laevis* and *Xenopus mulleri, Cold Spring Harbor Symp. Quant. Biol.,* **38,** 501–505.

Brown, D. D., P. C. Wensink, and E. Jordan, 1971, Purification and some characteristics of 5S DNA from *Xenopus laevis, Proc. Natl. Acad. Sci. U.S.A.,* **68,** 3175–3179.

Brown, D. D., P. C. Wensink, and E. Jordan, 1972, Comparison of the ribosomal DNA's of *Xenopus laevis* and *Xenopus mulleri:* the evolution of tandem genes, *J. Mol. Biol.,* **63,** 57–73.

Buettner-Janusch, J., V. Buettner-Janusch, and G. A. Mason, 1969, Amino acid compositions and amino-terminal end groups of alpha and beta chains from polymorphic hemoglobins of *Pongo pygmaeus, Arch. Biochem. Biophys.,* **133,** 164–170.

Burnham, C., 1962, *Discussions in Cytogenetics,* Burgess Publ. Co., Minneapolis.

Bush, G., 1969, Sympatric host race formation and speciation in frugivorous flies of the genus *Rhagoletis* (Diptera, Tephritidae), *Evolution,* **23,** 237–251.

Bush, G., 1974, The mechanism of sympatric host race formation in the true fruit flies (Tephritidae), in *Genetic Mechanisms of Speciation in Insects,* M. J. D. White (ed.), Australian and New Zealand Book Co., Sydney.

Bush, G. L., 1975, Modes of animal speciation, *Ann. Rev. Ecol. System.,* **6,** 339–364.

Calder, N., 1973, *The Life Game: Evolution and the New Biology,* Viking Press, New York.

Campbell, C. A., J. W. Valentine, and F. J. Ayala, High genetic variability in a population of *Tridacna maxima* from the Great Barrier Reef, *Mar. Biol.,* in press.

Campo, M. S., and J. O. Bishop, 1974, Two classes of messenger RNA in cultured rat cells: repetitive sequence transcripts and unique sequence transcripts, *J. Mol. Biol.,* **90,** 649–663.

Carlson, P., 1972, Locating genetic loci with amphiploids, *Mol. Gen. Genet.,* **114,** 273–280.

Carson, H. L., 1971, Speciation and the founder principle, *Stadler Symp.,* **3,** 51–70.

Carson, H. L., 1973, Reorganization of the gene pool during speciation, in *Genetics Structure of Populations,* N. E. Morton (ed.), *Pop. Genet. Monogr.,* **3,** 274–280.

Case, S. M., P. G. Haneline, and M. F. Smith, 1975, Protein variation in several species of *Hyla, System. Zool.,* **24,** 281–295.

Chamberlin, M. E., R. J. Britten, and E. H. Davidson, 1975, Sequence organization in *Xenopus* DNA studied by the electron microscope, *J. Mol. Biol.,* **96,** 317–333.

Champion, A. B., E. M. Prager, D. Wachter, and A. C. Wilson, 1974, Micro-complement fixation, in *Biochemical and Immunological Taxonomy of Animals,* C. A. Wright (ed.), Academic Press, London, pp. 397–416.

Champion, A. B., K. L. Soderberg, A. C. Wilson, and R. P. Ambler, 1975, Immunological comparison of azurins of known amino acid sequence: dependence of cross-reactivity upon sequence resemblance, *J. Mol. Evol.,* **5,** 291–305.

Chargaff, E., 1951, Structure and function of nucleic acids as cells constituents, *Fed. Proc.,* **10,** 654–659.

Chase, J. W., and C. C. Richardson, 1974*a*, Exonuclease VII of *Escherichia coli:* purification and properties, *J. Biol. Chem.,* **249,** 4545–4552.

Chase, J. W., and C. C. Richardson, 1974*b*, Exonuclease VII of *Escherichia coli:* mechanism of action, *J. Biol. Chem.,* **249,** 4553–4561.

Cherry, J. P., F. Katterman, and J. Endrizzi, 1970, Comparative studies of seed proteins of species of *Gossypium* by gel electrophoresis, *Evolution,* **24,** 431–447.

Chetverikov, S. S., 1926 (1959), On certain aspects of the evolutionary process from the standpoint of genetics, English translation: *Proc. Amer. Philos. Soc.,* **105,** 167–195.

Chothia, C., 1975, Structural invariants in protein folding, *Nature,* **254,** 304–306.

Cimino, M. C., and G. F. Bahr, 1974, The nuclear DNA content and chromatin ultrastructure of the coelacanth *Latimeria chalumnae, Exp. Cell Res.,* **88,** 263–272.

Clarke, B., 1975, The contribution of ecological genetics to evolutionary theory: detecting the direct effects of natural selection on particular polymorphic loci, *Genetics,* **79,** 101–113.

Clarkson, S. G., M. L. Birnstiel, and I. F. Purdom, 1973, Clustering of transfer RNA genes of *Xenopus laevis, J. Mol. Biol.,* **79,** 411–429.

Clausen, J., 1951, *Stages in the Evolution of Plant Species,* Cornell University Press, Ithaca, New York.

Cohen, P. T. W., G. S. Omenn, A. G. Motulsky, S. Chen, and E. R. Giblett, 1973, Restricted variation in glycolytic enzymes of human brain and erythrocytes, *Nature, New Biol.,* **241,** 229–233.

Colombera, D., 1974, Chromosome evolution in the phylum echinodermata, *Z. Zool. System. Evol.,* **12,** 299–308.

Commoner, B., 1964, Roles of deoxyribonucleic acid in inheritance, *Nature,* **202,** 960–968.

Connell, J. H., 1970, A predator-prey system in the marine intertidal region. I. *Balamus glandula* and several predatory species of *Thais, Ecol. Monogr.,* **40,** 49–78.

Connell, J. H., and E. Orias, 1964, The ecological regulation of species diversity, *Amer. Natur.,* **98,** 399–414.

Conner, W. G., R. Hinegardner, and K. Bachmann, 1972, Nuclear amounts in polychaete annelids, *Experientia,* **28,** 502–504.

Crow, J. F., and C. Denniston (eds.), 1974, *Genetic Distance,* Plenum Press, New York.

Darnell, J. E., and R. Balint, 1970, The distribution of rapidly hybridizing RNA sequences in heterogeneous nuclear RNA and mRNA from HeLa cells, *J. Cell Physiol.,* **76,** 349–356.

Davidson, E. H., and R. J. Britten, 1973, Organization, transcription, and regulation in the animal genome, *Quart. Rev. Biol.,* **48,** 565–613.

Davidson, E. H., G. A. Galau, R. C. Angerer, and R. J. Britten, 1975*a,* Comparative aspects of DNA organization in metazoa, *Chromosoma,* **51,** 253–259.

Davidson, E. H., B. R. Hough, C. S. Amenson, and R. J. Britten, 1973, General interspersion of repetitive with non-repetitive sequence elements in the DNA of *Xenopus, J. Mol. Biol.,* **77,** 1–23.

Davidson, E. H., B. R. Hough, W. H. Klein, and R. J. Britten, 1975*b,* Structural genes adjacent to interspersed repetitive DNA sequences, Cell, **4,** 217–238.

Dawid, I. B., D. D. Brown, and R. H. Reeder, 1970, Composition and structure of chromosomal and amplified ribosomal DNA's of *Xenopus laevis, J. Mol. Biol.,* **51,** 341–360.

Dayhoff, M. O., 1972, *Atlas of Protein Sequence and Structure,* vol. 5, National Biomedical Research Foundation, Silver Spring, Maryland.

Dayton, P. K., 1971, Competition, disturbance and community organization: the provision and subsequent utilization of space in a rocky intertidal community, *Ecol. Monogr.,* **41,** 351–389.

Dayton, P. K., and R. R. Hessler, 1972, Role of biological disturbance in maintaining diversity in the deep sea, *Deep-Sea Res.,* **19,** 199–208.

de Vries, H., 1901, *Die Mutationstheorie,* Veit, Leipzig.

Dickerson, R. E., 1971, The structure of cytochrome *c* and the rates of molecular evolution, *J. Mol. Evol.,* **1,** 26–45.

Dickerson, R. E., and I. Geis, 1969, *The Structure and Actions of Proteins,* Harper and Row, New York.

Dobzhansky, Th., 1935, A critique of the species concept in biology, *Philos. Sci.,* **2,** 344–355.

Dobzhansky, Th., 1936, Studies on hybrid sterility. II, *Genetics,* **21,** 113–135.

Dobzhansky, Th., 1937, *Genetics and the Origin of Species,* Columbia University Press, New York.

Dobzhansky, Th., 1950, Evolution in the tropics, *Amer. Sci.,* **38,** 209–221.

Dobzhansky, Th., 1970, *Genetics of the Evolutionary Process,* Columbia University Press, New York.

Dobzhansky, Th., 1972, Species of *Drosophila, Science,* **177,** 664–669.

Dobzhansky, Th., H. Levene, B. Spassky, and N. Spassky, 1959, Release of genetic variability through recombination. III, *Genetics,* **44,** 75–94.

Durham, J. W., 1971, The fossil record and the origin of the Deuterostomata, *Proc. North Amer. Paleontol. Conv.,* **2,** 1104–1132.

Ebeling, A. W., N. B. Atkin, and P. Y. Setzer, 1971, Genome sizes of teleostean fishes: increases in some deep-sea fishes, *Amer. Natur.,* **105,** 549–561.

Eck, R. V., and M. O. Dayhoff, 1966, Evolution of the structure of ferredoxin based on living relics of primitive amino acid sequences, *Science,* **152,** 363–366.

Efron, Y., M. Peleg, and A. Ashri, 1973, Alcohol dehydrogenase allozymes in the safflower genus *Carthamus* L, *Biochem. Genet.,* **9,** 299–308.

Endow, S., and J. G. Gall, 1975, Differential replication of satellite DNA in polyploid tissues of *Drosophila virilis, Chromosoma,* **50,** 175–192.

Estes, R., 1975, *Xenopus* from the talaeocene of Brazil and its zoogeographic importance, Nature, **254,** 48–50.

Fincham, J., 1969, Symposium talk, Eleventh International Botanical Congress, Seattle, Washington.

Fincham, J., 1972, Heterozygous advantage as a likely general basis for enzyme polymorphisms, *Heredity,* **28,** 387–391.

Fisher, A. G., 1960, Latitudinal variations in organic diversity, *Evolution,* **14,** 50–74.

Fisher, R. A., 1930, *The Genetical Theory of Natural Selection,* paperback ed. 1958, Dover, New York.

Fitch, W. M., 1971*a,* The non-identity of invariant positions in the cytochrome *c* of different species, *Biochem. Genet.,* **5,** 231–241.

Fitch, W. M., 1971*b,* Toward defining the course of evolution: minimum change for a specific tree topology, *System. Zool.,* **20,** 406–416.

Fitch, W. M., and E. Margoliash, 1967, The construction of phylogenetic trees, *Science,* **155,** 279–284.

Fitch, W. M., and E. Margoliash, 1970, The usefulness of amino acid and nucleotide sequences in evolutionary studies, *Evol. Biol.,* **4,** 67–109.

Ford, E., 1965, *Genetic Polymorphism,* M.I.T. Press, Cambridge, Massachusetts.

Forget, B. G., C. A. Marotta, S. M. Weissman, and M. Cohen-Solal, 1975, Nucleotide sequences of the 3′-terminal untranslated region of messenger RNA for human beta globin chain. *Proc. Natl. Acad. Sci. U.S.A.,* **72,** 3614–3618.

Forget, B. G., C. A. Marotta, S. M. Weissman, I. M. Verma, R. P. McCaffrey, and D. Baltimore, 1974, Nucleotide sequences of human globin messenger RNA, *Ann. N.Y. Acad. Sci.,* **241,** 290–309.

Garber, E., 1974, Enzymes as taxonomic and genetic tools in *Phaseolus* and *Aspergillus, Israel J. Med. Sci.,* **10,** 268–277.

Georgiev, G. P., A. J. Varshavsky, A. P. Ryskov, and R. B. Church, 1974, On the structural organization of the transcriptional unit in animal chromosomes, *Cold Spring Harbor Symp. Quant. Biol.,* **38,** 869–884.

Gibson, J., 1970, Enzyme flexibility in *Drosophila melanogaster, Nature,* **227,** 959–960.

Gillespie, J., 1974, The role of environmental grain in the maintenance of genetic variation, *Amer. Natur.,* **108,** 831–836.

Gillespie, J. H., and K. Kojima, 1968, The degree of polymorphism in enzymes involved in energy production compared to that in nonspecific enzymes in two *Drosophila ananassae* populations, *Proc. Natl. Acad. Sci. U.S.A.,* **61,** 582–585.

Gillespie, J. H., and C. H. Langley, 1974, A general model to account for enzyme variation in natural populations, *Genetics,* **76,** 837–884.

Glassman, E., 1965, Genetic regulation of xanthine dehydrogenase in *Drosophila melanogaster, Fed. Proc.,* **24,** 1243–1251.

Goldberg, R. B., W. R. Crain, J. V. Ruderman, G. P. Moore, T. R. Barnett, R. C. Higgins, R. A. Galfand, G. A. Galau, R. J. Britten, and E. H. Davidson, 1975, DNA sequence organization in the genomes of five marine invertebrates, *Chromosoma,* **51,** 225–251.

Goldberg, R. B., G. A. Galau, R. J. Britten, and E. H. Davidson, 1973, Sequence content of sea urchin embryo messenger RNA, *Proc. Natl. Acad. Sci. U.S.A.,* **70,** 3516–3520.

Gooch, J. L., and T. J. M. Schopf, 1973, Genetic variability in the deep sea: relation to environmental variability, *Evolution,* **26,** 545–552.

Goodman, M., 1961, The role of immunochemical differences in the phyletic development of human behavior, *Hum. Biol.,* **33,** 131–162.

Goodman, M., 1962, Evolution of the immunological species specificity of human serum proteins, *Hum. Biol.,* **34,** 104–150.

Goodman, M., 1963*a,* Serological analysis of the systematics of recent hominoids, *Hum. Biol.,* **35,** 377–436.

Goodman, M., 1963*b,* Man's place in the phylogeny of the primates as reflected in serum proteins, in *Classification and Human Evolution,* S. L. Washburn (ed.), Aldine, Chicago, pp. 204–234.

Goodman, M., 1973, The chronicle of primate phylogeny contained in proteins, *Symp. Zool. Soc. London,* **33,** 339–375.

Goodman, M., 1974, Biochemical evidence on hominid phylogeny, *Ann. Rev. Anthro.,* **3,** 203–228.

Goodman, M., W. Farris, Jr., G. W. Moore, W. Prychodko, E. Poulik, and M. W. Sorenson, 1974*a,* Immunodiffusion systematics of the primates. II. Findings on *Tarsius,* Lorisidae and Tupaiidae, in *Prosimian Biology,* R. D. Martin, G. A. Doyle, and A. C. Walker (eds.), Duckworth, London, pp. 881–890.

Goodman, M., and G. W. Moore, 1971, Immunodiffusion systematics of the primates. I. The Catarrhini, *System. Zool.,* **20,** 19–62.

Goodman M., G. W. Moore, J. Barnabas, and G. Matsuda, 1974*b,* The phylogeny of human globin genes investigated by the maximum method, *J. Mol. Evol.,* **3,** 1–48.

Goodman, M., G. W. Moore, and G. Matsuda, 1975, Darwinian evolution in the genealogy of haemoglobin, *Nature,* **253,** 603–608.

Gorham, S. W., 1974, *Checklist of World Amphibians up to January 1, 1970,* New Brunswick Museum, St. John, Canada.

Gorman, G. C., M. Soulé, S. Y. Yang, and E. Nevo, 1975, Evolutionary genetics of insular Adriatic lizards, *Evolution,* **29,** 52–71.

Gottlieb, L. D., 1972, A proposal for the classification of the annual species of *Stephanomeria* (Compositae), *Madrono,* **21,** 463–481.

Gottlieb, L. D., 1973*a,* Genetic differentiation, sympatric speciation and the origin of a diploid species of *Stephanomeria, Amer. J. Bot.,* **60,** 545–553.

Gottlieb, L. D., 1973*b,* Enzyme differentiation and phylogeny in *Clarkia franciscana, C. rubicunda* and *C. amoena, Evolution,* **27,** 205–214.

Gottlieb, L. D., 1973*c,* Genetic control of glutamate oxaloacetate transaminase in the diploid plant *Stephanomeria exigua* and its allotetraploid derivative, *Biochem. Genet.,* **9,** 97–107.

Gottlieb, L. D., 1974*a,* Gene duplication and fixed heterozygosity for alcohol dehydrogenase in the diploid plant *Clarkia franciscana, Proc. Natl. Acad. Sci. U.S.A.,* **71,** 1816–1818.

Gottlieb, L. D., 1974*b,* Genetic confirmation of the origin of *Clarkia lingulata, Evolution,* **28,** 244–250.

Gottlieb, L. D., 1974*c,* Genetic stability in a peripheral isolate of *Stephanomeria exigua* ssp. *coronaria* that fluctuates in population size, *Genetics,* **76,** 551–556.

Gottlieb, L. D., and G. Pilz, 1975, Genetic variation in the self-incompatible *Gaura demareei,* a recent derivative species, in MS.

Gould, S. J., 1975, This view of life. The child as man's real father, *Natur. Hist.,* **84,** 18–22.

Gould, S. J., D. S. Woodruff, and J. P. Martin, 1974, Genetics and morphometrics of *Cerion* at Pongo Carpet: a new systematic approach to this enigmatic land snail, *System. Zool.,* **23,** 518–535.

Graham, D. E., B. R. Neufeld, E. H. Davidson, and R. J. Britten, 1974, Interspersion of repetitive and non-repetitive DNA sequences in the sea urchin genome, *Cell,* **1,** 127–137.

Grant, V., 1964, The biological composition of a taxonomic species in *Gilia, Adv. Genet.,* **12,** 281–328.

Grant, V., 1966, The selective origin of incompatibility barriers in the plant genus *Gilia, Amer. Natur.,* **100,** 99–118.

Grant, V., and A. Grant, 1960, Genetic and taxonomic studies in *Gilia.* XI. Fertility relationships of the diploid Cobwebby Gilias, *Aliso,* **4,** 435–481.

Grassle, J. F., 1972, Species diversity, genetic variability and environmental uncertainty, *Fifth European Mar. Biol. Symp.,* Piccin, Padua, pp. 19–26.

Grassle, J. F., 1973, Variety in coral reef communities, in *Biology and Geology of Coral Reefs,* O. A. Jones and R. Endean (eds.), *Biology,* **1,** 247–270, Academic Press, New York.

Grassle, J. F., and H. L. Sanders, 1973, Life histories and the role of disturbance, *Deep-Sea Res.,* **20,** 643–659.

Hall, W. P., and R. K. Selander, 1973, Hybridization of karyotypically differentiated populations in the *Sceloporus grammicus* complex (Iguanidae), *Evolution,* **27,** 226–242.

Harris, H., 1966, Enzyme polymorphisms in man, *Proc. Roy. Soc. Ser. B.,* **164,** 298–310.

Harris, H., and D. A. Hopkinson, 1972, Average heterozygosity per locus in man: an estimate based on the incidence of enzyme polymorphism, *Ann. Hum. Genet.,* **36,** 9–20.

Harris, H., D. A. Hopkinson, and E. B. Robson, 1974, The incidence of rare alleles determining electrophoretic variants: data on 43 enzyme loci in man, *Ann. Hum. Genet.,* **37,** 237–253.

Hart, G., 1969, Genetic control of alcohol dehydrogenase isozymes in *Triticum dicoccum, Biochem. Genet.,* **3,** 617–625.

Hartigan, J. A., 1973, Minimum mutation fits to a given tree, *Biometrics,* **29,** 53–65.

Hartl, D. L., 1971, Some aspects of natural selection in arrhenotokous populations, *Amer. Zool.,* **11,** 309–325.

Hedgecock, D., 1974, Protein variation and evolution in the genus *Taricha* (Salamandridae), Ph.D. dissertation, University of California, Davis.

Hedgecock, D., and F. J. Ayala, 1974, Evolutionary divergence in the genus *Taricha* (Salamandridae), *Copeia,* **1974,** 738–747.

Hendrickson, W. A., and W. E. Love, 1971, Structure of lamprey haemoglobin, *Nature, New Biol.,* **232,** 197–203.

Hershey, A. D., and M. Chase, 1952, Independent functions of viral protein and nucleic acid in growth of bacteriophage, *J. Genet. Physiol.,* **36,** 39–56.

Hessler, R. R., and P. A. Jumars, 1974, Abyssal community analysis from replicate box cores in the central North Pacific, *Deep-Sea Res.,* **21,** 185–209.

Hessler, R. R., and H. L. Sanders, 1967, Faunal diversity in the deep-sea, *Deep-Sea Res.,* **14,** 65–78.

Hinegardner, R., 1968, Evolution of cellular DNA content in teleost fishes, *Amer. Natur.,* **102,** 517–523.

Hinegardner, R., 1974*a,* Cellular DNA content of the mollusca, *Comp. Biochem. Physiol.,* **47A,** 447–460.

Hinegardner, R., 1974*b,* Cellular DNA content of the echinodermata, *Comp. Biochem. Physiol.,* **49B,** 219–226.

Hinegardner, R., and D. E. Rosen, 1972, Cellular DNA content and the evolution of teleost fishes, *Amer. Natur.,* **106,** 621–644.

Hochachka, P., and G. Somero, 1973, *Strategies of Biochemical Adaptation,* Saunders, Philadelphia.

Holmes, D. S., and J. Bonner, 1974, Interspersion of repetitive and single-copy sequences in nuclear ribonucleic acid of high molecular weight, *Proc. Natl. Acad. Sci. U.S.A.,* **71,** 1108–1112.

Holmes, D. S., J. E. Mayfield, and J. Bonner, 1974, Sequence composition of rat ascites chromosomal ribonucleic acid, *Biochemistry,* **13,** 849–855.

Holmquist, R., 1972, Empirical support for a stochastic model of evolution, *J. Mol. Evol.,* **1,** 211–222.

Hood, L., J. Campbell, and S. Elgin, 1975, The organization, expression and evolution of antibodies and other multigene families, *Ann. Rev. Genet.,* **9,** 305–353.

Hough, B. R., and E. H. Davidson, 1972, Studies on the repetitive sequence transcripts of *Xenopus* oocytes, *J. Mol. Biol.,* **70,** 491–509.

Hough, B. R., M. J. Smith, R. J. Britten, and E. H. Davidson, 1975, Sequence complexity of heterogeneous nuclear RNA in sea urchin embryos, Cell, **5,** 291–299.

Hoyer, B. H., and R. B. Roberts, 1967, Studies of nucleic acid interactions using DNA-agar, in *Molecular Genetics,* pt. II, H. Taylor (ed.), Academic Press, New York, pp. 425–479.

Hoyer, B. H., N. W. van de Velde, M. Goodman, and R. B. Roberts, 1972, Examination of hominid evolution by DNA sequence homology, *J. Hum. Evol.,* **1,** 645–649.

Hunt, W. G., and R. K. Selander, 1973, Biochemical genetics of hybridization in European house mice, *Heredity,* **31,** 11–33.

Hunter, R. L., and C. L. Markert, 1957, Histochemical demonstration of enzymes separated by zone electrophoresis in starch gels, *Science,* **125,** 1294–1295.

Inger, R. F., H. K. Voris, and H. H. Voris, 1974, Genetic variation and population ecology of some Southeast Asian frogs of the genera *Bufo* and *Rana, Biol. Genet.,* **12,** 121–145.

Ingram, V., 1957, Gene mutations in normal haemoglobin: the chemical difference between normal and sickle cell haemoglobin, *Nature,* **180,** 326–328.

Jacob, F., and E. L. Wollman, 1961, *Sexuality and Genetics of Bacteria,* Academic Press, New York.

Jelinek, W., G. Molloy, M. Salditt, R. Wall, D. Sheiness, and J. E. Darnell, Jr., 1974, Origin of mRNA in HeLa cells and the implications for chromosome structure, *Cold Spring Harbor Symp. Quant. Biol.,* **38,** 891–898.

Johnson, B. L., and O. Hall, 1965, Analysis of phylogenetic affinities in the Triticinae by protein electrophoresis, *Amer. J. Bot.,* **57,** 977–987.

Johnson, G. B., 1971, Metabolic implications of polymorphism as an adaptive strategy, *Nature,* **232,** 347–349.

Johnson, G. B., 1972, The selective significance of biochemical polymorphisms in *Colias* butterflies, Ph.D. dissertation, Stanford University.

Johnson, G. B., 1973, Importance of substrate variability to enzyme polymorphisms, *Nature, New Biol.,* **243,** 151–153.

Johnson, G. B., 1974*a,* On the estimation of effective number of alleles from electrophoretic data, *Genetics,* **78,** 771–776.

Johnson, G. B., 1974*b,* Enzyme polymorphism and metabolism, *Science,* **184,** 28–37.

Johnson, G. B., 1975, Enzyme polymorphism and adaptation, *Stadler Genet. Symp.,* **7,** 91–116.

Johnson, R. G., 1970, Variations in diversity within benthic marine communities, *Amer. Natur.,* **104,** 285–300.

Johnson, W. E., H. L. Carson, K. Y. Kaneshiro, W. W. M. Steiner, and M. M. Cooper, 1975, Genetic variation in Hawaiian *Drosophila.* II. Allozymic differentiation in the *D. planitibia* subgroup, in *Isozymes. IV. Genetics and Evolution,* C. L. Markert (ed.), Academic Press, New York, pp. 563–584.

Johnson, W. E., and R. K. Selander, 1971, Protein variation and systematics in kangaroo rats (genus *Dipodomys*), *System. Zool.,* **20,** 377–405.

Jukes, T. H., and R. Holmquist, 1972, Evolutionary clock: nonconstancy of rate in different species, *Science,* **177,** 530-532.

Kauzmann, W., 1959, Some factors in the interpretation of protein denaturation, *Adv. Protein Chem.,* **14,** 1.

Kedes, L. H., and M. L. Birnstiel, 1971, Reiteration and clustering of DNA sequences complementary to histone messenger RNA, *Nature, New Biol.,* **230,** 165–169.

Kimura, M., 1968, Evolutionary rate at the molecular level, *Nature,* **217,** 624–626.

Kimura, M., 1969, The rate of molecular evolution considered from the standpoint of population genetics, *Proc. Natl. Acad. Sci. U.S.A.,* **63,** 1181–1188.

Kimura, M., 1971, Theoretical foundations of population genetics at the molecular level, *Theor. Pop. Biol.,* **2,** 174–208.

Kimura, M., and T. Maruyama, 1971, Pattern of neutral polymorphism in a geographically structured population, *Genet. Res.,* **18,** 125–131.

Kimura, M., and T. Ohta, 1971*a,* Protein polymorphism as a phase of molecular evolution, *Nature,* **229,** 467–469.

Kimura, M., and T. Ohta, 1971*b*, *Theoretical Aspects of Population Genetics,* Princeton University Press, Princeton, New Jersey.

King, J. L., 1973, The probability of electrophoretic identity of proteins as a function of amino acid divergence, *J. Mol. Evol.,* **2,** 317–322.

King, J. L., 1974, Isoallele frequencies in very large populations, *Genetics,* **76,** 607–613.

King, J. L., 1975, Review of "The Genetic Basis of Evolutionary Change," by R. C. Lewontin, *Ann. Hum. Genet.,* **38,** 507–511.

King, J. L., and T. H. Jukes, 1969, Non-Darwinian evolution: random fixation for selectively neutral alleles, *Science,* **164,** 788–798.

King, J. L., and T. H. Ohta, 1975, Polyallelic mutational equilibria, *Genetics,* **79,** 681–691.

King, M. C., and A. C. Wilson, 1975, Evolution at two levels. Molecular similarities and biological differences between humans and chimpanzees, *Science,* **188,** 107–116.

Klein, W. H., W. Murphy, G. Attardi, R. J. Britten, and E. H. Davidson, 1974, Distribution of repetitive and nonrepetitive sequence transcripts in HeLa mRNA, *Proc. Natl. Acad. Sci. U.S.A.,* **71,** 1785–1789.

Klopfer, P. H., 1959, Environmental determinants of faunal diversity, *Amer. Natur.,* **93,** 337–342.

Klotz, I. M., 1967, Protein subunits: a table, *Science,* **155,** 697–698.

Kohne, D. E., 1970, Evolution of higher-organism DNA, *Quart. Rev. Biophys.,* **3,** 327–375.

Kohne, D. E., J. A. Chiscon, and B. H. Hoyer, 1972, Evolution of primate DNA sequences, *J. Hum. Evol.,* **1,** 627–644.

Kojima, K., J. Gillespie, and Y. N. Tobari, 1970, A profile of *Drosophila* species' enzymes assayed by electrophoresis. I. Number of alleles, heterozygosities, and linkage disequilibrium in glucose-metabolizing systems and some other enzymes, *Biochem. Genet.,* **4,** 627–637.

Kurten, B., 1968, *Pleistocene Mammals of Europe,* Aldine, Chicago.

Kyhos, D., 1965, The independent aneuploid origin of two species of *Chaenactis* (Compositae) from a common ancestor, *Evolution,* **19,** 26–43.

Laird, C. D., 1973, DNA of *Drosophila* chromosomes, *Ann. Rev. Genet.,* **7,** 177–204.

Laird, C. D., B. L. McConaughy, and B. J. McCarthy, 1969, Rate of fixation of nucleotide substitutions in evolution, *Nature,* **224,** 149–154.

Lane, C. D., G. Marbaix, and J. B. Gurdon, 1971, Rabbit haemoglobin synthesis in frog cells: the translation of reticulocyte 9S RNA in frog oocytes, *J. Mol. Biol.,* **61,** 73–91.

Langley, C. H., and W. M. Fitch, 1973, The constancy of evolution: a statistical analysis of the α and β hemoglobins, cytochrome *c* and fibrinopeptide A, in *Genetic Structure of Populations,* N. E. Morton (ed.), University Press Hawaii, Honolulu, pp. 246–262.

Langley, C. H., and W. M. Fitch, 1974, An examination of the constancy of the rate of molecular evolution, *J. Mol. Evol.,* **3,** 161–177.

Le Cam, L. M., J. Neyman, and E. L. Scott (eds.), 1972, *Darwinian, Neo-Darwinian, and Non-Darwinian Evolution, Proc. Sixth Berkeley Symp. Math. Stat. Prob.,* vol. 5, University of California Press, Berkeley.

Lerner, S. A., T. T. Wu, and E. C. C. Lin, 1964, Evolution of a catabolic pathway in bacteria, *Science,* **146,** 1313–1315.

Lester, L. J., 1975, Population genetics of the Hymenoptera, Ph.D. dissertation, University of Texas, Austin.

Levene, H., 1953, Genetic equilibrium when more than one ecological niche is available, *Amer. Natur.,* **87,** 331–333.

Levin, D., 1968, The genome constitution of eastern North American *Phlox* amphiploids, *Evolution,* **22,** 612–632.

Levin, D., and B. Schaal, 1970, Reticulate evolution in *Phlox* as seen through protein electrophoresis, *Amer. J. Bot.,* **57,** 977–987.

Levin, D. A., 1975, Genic heterozygosity and protein polymorphism among local populations of *Oenothera biennis, Genetics,* **79,** 477–491.

Levins, R., 1968, *Evolution in Changing Environments,* Princeton University Press, Princeton, New Jersey.

Levins, R., and R. MacArthur, 1966, The maintenance of genetic polymorphism in a spatially heterogeneous environment: variations on a theme by Howard Levene, *Amer. Natur.,* **100,** 585–589.

Levy, M., and D. Levin, 1971, The origin of novel flavonoids in *Phlox* allotetraploids, *Proc. Natl. Acad. Sci. U.S.A.,* **68,** 1627–1630.

Levy, M., and D. Levin, 1974, Novel flavonoids and reticulate evolution in the *Phlox pilosa–P. drummondii* complex, *Amer. J. Bot.,* **61,** 156–167.

Lewin, B., 1975, Units of transcription and translation: sequence components of heterogeneous nuclear RNA and messenger RNA, *Cell,* **4,** 77–93.

Lewis, H., 1962, Catastrophic selection as a factor in speciation, *Evolution,* **16,** 257–271.

Lewis, H., 1966, Speciation in flowering plants, *Science,* **152,** 167–172.

Lewis, H., 1973, The origin of diploid neospecies in *Clarkia, Amer. Natur.,* **107,** 161–170.

Lewis, H., and M. Roberts, 1956, The origin of *Clarkia lingulata, Evolution,* **10,** 126–138.

Lewontin, R. C., 1967, An estimate of average heterozygosity in man, *Amer. J. Hum. Genet.,* **19,** 681–685.

Lewontin, R. C., 1974, *The Genetic Basis of Evolutionary Change,* Columbia University Press, New York.

Lewontin, R. C., and J. L. Hubby, 1966, A molecular approach to the study of genic heterozygosity in natural populations. II. Amount of variation and degree of heterozygosity in natural populations of *Drosophila pseudoobscura*, *Genetics*, **54**, 595–609.

Li, C. H., 1972, Adrenocorticotropin 45. Revised amino acid sequences for sheep and bovine hormones, *Biochem. Biophys. Res. Comm.*, **49**, 835–839.

Lin, C. C., B. Chiarelli, L. E. M. de Boer, and M. M. Cohen, 1973, A comparison of the fluorescent karyotypes of the chimpanzee (*Pan troglodytes*) and man, *J. Hum. Evol.*, **2**, 311–321.

Lloyd, D., 1965, Evolution of self-compatibility and racial differentiation in *Leavenworthia* (Cruciferae), *Contr. Gray Herb. Harv.*, **195**, 1–134.

Lokki, J., E. Suomalainen, A. Saura, and P. Lankinen, 1975, Genetic polymorphism and evolution in parthenogenetic animals. II. Diploid and polyploid *Solenobia triquetrella* (Lepidoptera: Psychidae), *Genetics*, **70**, 513–525.

Lotsy, J. P., 1916, *Evolution by Means of Hybridization*, Nijhoff, The Hague.

Ludwig, W., 1950, Zur Theorie der Konkurenz. Die Annidation (Einnischung) als fünfter Evolutionsfaktor. Neue Ergeb, *Probleme Zool., Klatt-Festschrift*, **1950**, 516–537.

MacDonald, T., and J. Brewbaker, 1974, Isoenzyme polymorphism in flowering plants. IX. The E5-E10 esterase loci of maize, *J. Hered.*, **65**, 37–42.

Maisel, H., 1965, Phylogenetic properties of primate lens antigens, in *Protides of the Biological Fluids—1964*, H. Peeters (ed.), Elsevier, Amsterdam, pp. 146–148.

Manning, J. E., C. W. Schmid, and N. Davidson, 1975, Interspersion of repetitive and nonrepetitive DNA sequences in the *Drosophila melanogaster* genome, *Cell*, **4**, 144–155.

Manwell, C., and C. M. Baker, 1970, *Molecular Biology and the Origin of Species*, University of Washington Press, Seattle.

Mao, S., and H. C. Dessauer, 1971, Selectively neutral mutations, transferrins and the evolution of natricine snakes, *Comp. Biochem. Physiol.*, **40A**, 669–680.

Margalef, R., 1968, *Perspectives in Ecological Theory*, University of Chicago Press.

Margoliash, E., and W. M. Fitch, 1968, Evolutionary variability of cytochrome *c* primary structures, *Ann. N.Y. Acad. Sci.*, **151**, 359–381.

Margoliash, E., W. M. Fitch, and R. E. Dickerson, 1971, Molecular expression of evolutionary phenomena in the primary and tertiary structures of cytochrome *c*, in *Biochemical Evolution and the Origin of Life*, E. Schoffeniels (ed.), North-Holland/American Elsevier, New York, pp. 52–95.

Margulis, L., 1974, Five kingdom classification and the origin and evolution of cells, *Evol. Biol.*, **7**, 45–78.

Marinković, D., and F. J. Ayala, 1975, Fitness of allozyme variants in *Drosophila pseudoobscura*. II. Selection at the *Est-5*, *Odh* and *Mdh-2* loci, *Genet. Res.*, **24**, 137–149.

Markert, C. L., 1968, The molecular basis for isozymes, *Ann. N.Y. Acad. Sci.*, **151**, 14–40.

Markert, C. L. (ed.), 1975, *Isozymes,* vol. I to IV, Academic Press, New York.

Markert, C. L., and F. Møller, 1959, Multiple forms of enzymes: tissue, ontogenetic, and species specific patterns, *Proc. Natl. Acad. Sci. U.S.A.,* **45,** 753–763.

Maruyama, T., and M. Kimura, 1974, Geographical uniformity of selectively neutral polymorphisms, *Nature,* **249,** 30–32.

Mauridis, A., A. Tulinsky, and M. Liebman, 1974, Asymmetrical changes in the tertiary structure of α-chymotrypsin with change in pH, *Biochemistry,* **13,** 3661–3666.

Maxson, L. R., V. M. Sarich, and A. C. Wilson, 1975, Continental drift and the use of albumin as an evolutionary clock, *Nature,* **225,** 397–400.

Maxson, L. R., and A. C. Wilson, 1974, Convergent morphological evolution detected by studying proteins of tree frogs in the *Hyla eximia* species group, *Science,* **185,** 66–68.

Maxson, L. R., and A. C. Wilson, 1975, Albumin evolution and organismal evolution in tree frogs (Hylidae), *System. Zool.,* **24,** 1–15.

Mayr, E., 1942, *Systematics and the Origin of Species,* Columbia University Press, New York.

Mayr, E., 1963, *Animal Species and Evolution,* Harvard University Press, Cambridge, Massachusetts.

Mayr, E., 1970, *Population, Species, and Evolution,* Belknap Press, Cambridge, Massachusetts.

McDonald, J. F., and F. J. Ayala, 1974, Genetic response to environmental heterogeneity, *Nature,* **250,** 572–574.

McLachlan, A. D., 1972, Gene duplication in carp muscle calcium binding protein, *Nature,* **240,** 83–85.

Merritt, R., 1972, Geographic distribution and enzymatic properties of lactate dehydrogenase allozymes in the fathead minnow, *Amer. Natur.,* **106,** 173–184.

Milkman, R., 1975, Allozyme variation in *E. coli* of diverse natural origins, in *Isozymes. IV. Genetics and Evolution,* C. L. Markert (ed.), Academic Press, New York, pp. 273–285.

Mirsky, A. E., and H. Ris, 1949, Variable and constant components of chromosomes, *Nature,* **163,** 666–667.

Mirsky, A. E., and H. Ris, 1951, The deoxyribonucleic acid content of animal cells and its evolutionary significance, *J. Genet. Physiol.* **34,** 451–462.

Mitra, R., and C. Bhatia, 1971, Isoenzymes and polyploidy. I. Qualitative and quantitative isoenzyme studies in the Triticinae, *Genet. Res., Camb.,* **18,** 57–69.

Mizuno, S., and H. C. Macgregor, 1974, Chromosomes, DNA sequences, and evolution in salamanders of the genus *Plethodon, Chromosoma,* **48,** 239–296.

Molloy, G. R., W. Jelinek, M. Salditt, and J. E. Darnell, 1974, Arrangement of specific oligonucleotides within poly(A) terminated hnRNA molecules, *Cell,* **1,** 43–53.

Moore, G. W., 1975, Proof of the maximum parsimony algorithm, paper prepared for *Burg Wartenstein Symposium No. 65, Progress in Molecular Anthropology.*

Moore, G. W., J. Barnabas, and M. Goodman, 1973, A method for constructing maximum parsimony ancestral amino acid sequences on a given network, *J. Theor. Biol.,* **38,** 459–485.

Moore, G. W., and M. Goodman, 1968, A set theoretical approach to immunotaxonomy: analysis of species comparisons in modified Ouchterlony plates, *Bull. Math. Biophys.,* **30,** 279–289.

Morgan, T. H., 1919, *The Physical Basis of Heredity,* Lippincott, Philadelphia.

Muller, H. J., 1950, Evidence of the precision of genetic adaptation, *Harvey Lectures,* **43,** 165–229.

Muller, H. J., and W. D. Kaplan, 1966, The dosage compensation of *Drosophila* and mammals as showing the accuracy of the normal type, *Genet. Res.,* **8,** 41–59.

Murray, B., and C. Williams, 1973, Polyploidy and flavonoid synthesis in *Briza media* L, *Nature,* **243,** 87–88.

Nei, M., 1972, Genetic distance between populations, *Amer. Natur.,* **106,** 283–292.

Nei, M., 1975, *Molecular Population Genetics and Evolution,* North-Holland, Amsterdam.

Nei, M., and R. Chakraborty, 1973, Genetics distance and electrophoretic identity of proteins between taxa, *J. Mol. Evol.,* **2,** 323–328.

Nei, M., T. Maruyama, and R. Chakraborty, 1975, The bottleneck effect and genetic variability in populations, *Evolution,* **29,** 1–10.

Nei, M., and A. K. Roychoudhury, 1974, Genic variation within and between the three major races of man, Caucasoids, Negroids, and Mongoloids, *Amer. J. Hum. Genet.,* **26,** 421–443.

Neurath, H., K. A. Walsh, and W. P. Winter, 1967, Evolution of structure and function of proteases, *Science,* **158,** 1638–1644.

Nevo, E., Y. J. Kim, C. R. Shaw, and C. S. Thaeler, 1974, Genetic variation, selection, and speciation in *Thomomys talpoides* pocket gophers, *Evolution,* **28,** 1–23.

Nevo, E., and C. R. Shaw, 1972, Genetic variation in a subterranean mammal, *Spalax ehrenbergi, Biochem. Genet.,* **7,** 235–241.

Nolan, C., and E. Margoliash, 1968, Comparative aspects of primary structures of proteins, *Ann. Rev. Biochem.,* **37,** 727–790.

Nolan, R. A., A. H. Brush, N. Arnheim, and A. C. Wilson, 1975, An inconsistency between protein resemblance and taxonomic resemblance: immunological comparison of diverse proteins from gallinaceous birds, *Condor,* **77,** 154–159.

Nozawa, K., T. Shotake, and Y. Okura, 1975, Blood protein polymorphisms and population structure of the Japanese macaque, *Macaca fuscata fuscata,* in *Isozymes. IV. Genetics and Evolution,* C. L. Markert (ed.), Academic Press, New York, pp. 225–241.

Nuttall, G. H. F., 1904, *Blood Immunity and Blood Relationship,* Cambridge University Press, Cambridge, England.

Ohno, S., 1969, The preferential activation of maternally derived alleles in development of interspecific hybrids, in *Heterospecific Genome Interaction,* V. Defendi (ed.), Wistar Instit. Press, Philadelphia, pp. 137–150.

Ohno, S., 1970, *Evolution by Gene Duplication,* Springer-Verlag, New York.

Ohta, T., 1974, Mutational pressure as the main cause of molecular evolution and polymorphism, *Nature,* **252,** 351–354.

Ohta, T., and M. Kimura, 1971, On the constancy of the evolutionary rate of cistrons, *J. Mol. Evol.,* **1,** 18–25.

Ohta, T., and M. Kimura, 1973, A model of mutation appropriate to estimate the number of electrophoretically detectable alleles in a finite population, *Genet. Res.,* **22,** 201–204.

Ohta, T., and M. Kimura, 1974, Simulation studies on electrophoretically detectable genetic variability in a finite population, *Genetics,* **76,** 615–624.

Ohta, T., and M. Kimura, 1975, Theoretical analysis of electrophoretically detectable polymorphisms: models of very slightly deleterious mutations, *Amer. Natur.,* **109,** 137–145.

Olmo, E., 1973, Quantitative variations in the nuclear DNA and phylogenesis of the amphibia, *Caryologia,* **26,** 43–68.

Omenn, G. S., P. T. W. Cohen, and A. G. Motulsky, 1971, Genetic variation in glycolytic enzymes in human brain, *Excerpta Med. Intern. Cong. Ser.,* **233,** 135.

Ownbey, M., 1950, Natural hybridization and amphiploidy in the genus *Tragopogon, Amer. J. Bot.,* **37,** 487–499.

Paine, R. T., 1966, Food web complexity and species diversity, *Amer. Natur.,* **100,** 65–75.

Pedersen, R. A., 1971, DNA content, ribosomal gene multiplicity, and cell size in fish, *J. Exp. Zool.,* **177,** 65–78.

Perkins, D. D., 1974, The manifestation of chromosome rearrangements in unordered asci of *Neurospora, Genetics,* **77,** 459–489.

Perutz, M. F., and L. F. Ten Eyck, 1972, Stereochemistry of cooperative effects in hemoglobin, *Cold Spring Harbor Symp. Quant. Biol.,* **36,** 295–310.

Pianka, E. R., 1966, Latitudinal gradients in species diversity: a review of concepts, *Amer. Natur.,* **100,** 33–46.

Pontecorvo, G., 1943, Viability interactions between chromosomes of *Drosophila melanogaster* and *Drosophila simulans, J. Genet.,* **43,** 51–66.

Powell, J. R., 1971, Genetic polymorphisms in varied environments, *Science,* **174,** 1035–1036.

Prager, E. M., N. Arnheim, G. A. Mross, and A. C. Wilson, 1972, Amino acid sequence studies on bobwhite quail egg white lysozyme, *J. Biol. Chem.*, **247,** 2905–2916.

Prager, E. M., and A. C. Wilson, 1971, The dependence of immunological cross-reactivity upon sequence resemblance among lysozymes, *J. Biol. Chem.*, **246,** 5978–5989.

Prager, E. M., and A. C. Wilson, 1975, Slow evolutionary loss of the potential for interspecific hybridization in birds. A manifestation of slow regulatory evolution, *Proc. Natl. Acad. Sci. U.S.A.*, **71,** 200–204.

Prakash, S., 1972, Origin of reproductive isolation in the absence of apparent genic differentiation in a geographic isolate of *Drosophila pseudoobscura*, *Genetics*, **72,** 143–155.

Prakash, S., 1973, Patterns of gene variation in central and marginal populations of *Drosophila robusta*, *Genetics*, **75,** 347–369.

Prakash, S., R. C. Lewontin, and J. L. Hubby, 1969, A molecular approach to the study of genic heterozygosity in natural populations. IV. Patterns of genic variation in central, marginal and isolated populations of *Drosophila pseudoobscura*, *Genetics*, **61,** 841–858.

Raup, D. M., S. J. Gould, T. J. M. Schopf, and D. S. Simberloff, 1973, Stochastic models of phylogeny and the evolution of diversity, *J. Geol.*, **81,** 525–542.

Raven, P., and D. Gregory, 1972, A revision of the genus *Gaura* (Onagraceae), *Mem. Torrey Bot. Club*, **23,** 1–96.

Reddy, M., and E. Garber, 1971, Genetic studies of variant enzymes. III. Comparative electrophoretic studies of esterases and peroxidases for species, hybrids and amphiploids in the genus *Nicotiana*, *Bot. Gaz.*, **132,** 158.

Rees, H., 1974, DNA in higher plants, in *Evolution of Genetics Systems*, H. H. Smith *et al.* (eds.), *Brookhaven Symp. Biol.*, **23,** 394–418.

Rensch, B., 1959, *Evolution above the Species Level*, Columbia University Press, New York.

Rheinsmith, E. L., R. Hinegardner, and K. Bachmann, 1974, Nuclear DNA amounts in crustacea, *Comp. Biochem. Physiol.*, **48B,** 343–348.

Rice, N., 1971, Thermal stability of reassociated repeated DNA from rodents, *Carnegie Inst. Wash. Year Book*, **69,** 472–479.

Rice, N., 1972, Changes in repeated DNA in evolution, in *Evolution of Genetic Systems*, H. H. Smith (ed.), *Brookhaven Symp.*, **23,** 44–78, Gordon and Breach, New York.

Rice, N. R., and N. A. Straus, 1973, Relatedness of mouse satellite DNA to DNA of various *Mus* species, *Proc. Natl. Acad. Sci. U.S.A.*, **70,** 3546–3550.

Richardson, R. H., M. E. Richardson, and P. E. Smouse, 1975, Evolution of electrophoretic mobility in the *Drosophila mulleri* complex, in *Isozymes. IV. Genetics and Evolution*, C. L. Markert (ed.), Academic Press, New York, pp. 533–545.

Rick, C., 1963, Barriers to interbreeding in *Lycopersicon peruvianum*, *Evolution*, **17,** 216–232.

Rick, C., and L. Butler, 1956, Cytogenetics of the tomato, *Adv. Genet.,* **8,** 267–282.

Rick, C. M., and P. G. Smith, 1953, Novel variation in tomato species hybrids, *Amer. Natur.,* **87,** 359–373.

Ritossa, F. M., K. C. Atwood, and S. Spiegelman, 1966, A molecular explanation of the *bobbed* mutants of *Drosophila* as partial deficiencies of ribosomal DNA, *Genetics,* **54,** 819–834.

Ritossa, F. M., and S. Spiegelman, 1965, Localization of DNA complementary to ribosomal RNA in the nucleolus organizer region of *Drosophila melanogaster, Proc. Natl. Acad. Sci. U.S.A.,* **53,** 737–745.

Rogers, J. S., 1972, Measures of genetic similarity and genetic distance, *Univ. Texas Publ.,* **7213,** 145–153.

Roose, M., and L. D. Gottlieb, 1975, Genetic consequences of polyploidy in *Tragopogon,* in MS.

Rossmann, M. G., D. Moras, and K. W. Olsen, 1974, Chemical and biological evolution of a nucleotide-binding protein, *Nature,* **250,** 194–199.

Ryffel, G. U., and B. J. McCarthy, 1975, Polyadenylated RNA complementary to repetitive DNA in mouse L-cells, *Biochemistry,* **14,** 1385–1389.

Sanders, H. L., 1968, Marine benthic diversity: a comparative study, *Amer. Natur.,* **102,** 243–282.

Sarich, V. M., 1972*a,* Generation time and albumin evolution, *Biochem. Genet.,* **7,** 205–212.

Sarich, V. M., 1972*b,* On the nonidentity of several carnivore hemoglobins, *Biochem. Genet.,* **7,** 253–258.

Sarich, V. M., 1973, The giant panda is a bear, *Nature,* **245,** 218–220.

Sarich, V. M., and J. E. Cronin, 1974, Primate evolution at higher taxon levels: a molecular view, *Amer. J. Phys. Anthro.,* **41,** 502.

Sarich, V. M., and A. C. Wilson, 1967, Immunological time scale for hominoid evolution, *Science,* **158,** 1200–1203.

Sarich, V. M., and A. C. Wilson, 1973, Generation time and genomic evolution in primates, *Science,* **171,** 1144–1147.

Saura, A., O. Halkka, and J. Lokki, 1973, Enzyme gene heterozygosity in small island populations of *Philaenus spumarius* (L.) (Homoptera), *Genetica,* **44,** 459–473.

Scandalios, J., 1974, Isozymes in development and differentiation, *Ann. Rev. Plant Physiol.,* **25,** 225–258.

Schopf, T. J. M., and L. S. Murphy, 1973, Protein polymorphism of the hybridizing sea stars *Asterias forbesi* and *Asterias vulgaris* and implications for their evolution, *Biol. Bull.,* **145,** 589–597.

Schwartz, D., 1973, Single gene heterosis for alcohol dehydrogenase in maize: the nature of the subunit interaction, *Theor. App. Genet.,* **43,** 117.

Schwartz, D., and T. Endo, 1966, Alcohol dehydrogenase polymorphism in maize—simple and compound loci, *Genetics,* **53,** 709–715.

Schwartz, D., and W. Laughner, 1969, A molecular basis for heterosis, *Science,* **166,** 626.

Selander, R. K., 1975, Stochastic factors in the genetic structure of populations, *Proc. Eight Intern. Conf. Numerical Taxonomy,* W. H. Freeman, San Francisco, in press.

Selander, R. K., and R. O. Hudson, 1975, Animal population structure under close inbreeding: the land snail *Rumina* in southern France, *Amer. Natur.,* in press.

Selander, R. K., W. G. Hunt, and S. Y. Yang, 1969, Protein polymorphism and genic heterozygosity in two European subspecies of the house mouse, *Evolution,* **23,** 379–390.

Selander, R. K., and W. E. Johnson, 1973, Genetic variation among vertebrate species, *Ann. Rev. Ecol. System.,* **4,** 75–91.

Selander, R. K., and D. W. Kaufman, 1973*a*, Genic variability and strategies of adaptation in animals, *Proc. Natl. Acad. Sci. U.S.A.,* **70,** 1875–1877.

Selander, R. K., and D. W. Kaufman, 1973*b*, Self-fertilization and genetic population structure in a colonizing land snail, *Proc. Natl. Acad. Sci. U.S.A.,* **70,** 1186–1190.

Selander, R. K., D. W. Kaufman, R. J. Baker, and S. L. Williams, 1975, Genic and chromosomal differentiation in pocket gophers of the *Geomys bursarius* group, *Evolution,* **28,** 557–564.

Selander, R. K., M. H. Smith, S. Y. Yang, W. E. Johnson, and J. B. Gentry, 1971, Biochemical polymorphism and systematics in the genus *Peromyscus.* I. Variation in the old-field mouse (*Peromyscus polionotus*), in *Stud. Genet. VI.* Univ. Texas Publ., **7103,** 49–90.

Selander, R. K., S. Y. Yang, R. C. Lewontin, and W. E. Johnson, 1970, Genetic variation in the horseshoe crab (*Limulus polyphemus*), a phylogenetic relic, *Evolution,* **24,** 402–414.

Sheen, S., 1972, Isozymic evidence bearing on the origin of *Nicotiana tabacum, Evolution,* **26,** 143–154.

Shields, G. F., and N. A. Straus, 1975, DNA–DNA hybridization studies of birds, *Evolution,* **29,** 159–166.

Shotton, D. M., and B. S. Hartley, 1970, Amino-acid sequence of porcine elastase and its homologies with other serine proteinases, *Nature,* **227,** 802–806.

Simpson, G. G., 1953, *The Major Features of Evolution,* Columbia University Press, New York.

Simpson, G. G., 1964, Species density of North American recent mammals, *System Zool.,* **13,** 57–73.

Sing, C. F., and A. R. Templeton, 1975, A search for the genetic unit of selection, in *Isozymes, IV. Genetics and Evolution,* C. L. Markert (ed.), Academic Press, New York, pp. 115–129.

Singh, R. S., 1975, Substrate specific enzyme variation in natural populations, *Genetics,* **80,** (suppl.), 75–76.

Small, E., 1971, The evolution of reproductive isolation in *Clarkia,* section Myxocarpa, *Evolution,* **25,** 330–346.

Smith, D., and D. Levin, 1963, Chromatographic study of reticulate evolution in the Appalachian Asplenium complex, *Amer. J. Bot.,* **50,** 952–958.

Smith, E. L., F. S. Markland, C. B. Kasper, R. J. DeLange, M. Landon, and W. H. Evans, 1966, The complete amino acid sequence of two types of subtilisin, BPN′ and Carlsberg, *J. Biol. Chem.,* **241,** 5974–5976.

Smith, G., 1974, Unequal crossover and the evolution of multigene families, *Cold Spring Harbor Symp. Quant. Biol.,* **38,** 507–513.

Smith, H. H., 1974, Evolution of genetic systems, *Brookhaven Symp. Biol.,* **23.**

Smith, H. H., D. Hamill, E. Weaver, and K. Thompson, 1970, Multiple molecular forms of peroxidases and esterases among *Nicotiana* species and amphiploids, *J. Hered.,* **61,** 203–212.

Smith, M. H., R. K. Selander, and W. E. Johnson, 1973, Biochemical polymorphism and systematics in the genus *Peromyscus.* III. Variation in the Florida deermouse (*Peromyscus floridanus*), a Pleistocene relic, *J. Mammal.,* **54,** 1–13.

Smith, M. J., B. R. Hough, M. E. Chamberlin, and E. H. Davidson, 1974, Repetitive and non-repetitive sequence in sea urchin heterogeneous nuclear RNA, *J. Mol. Biol.,* **85,** 103–126.

Sneath, P. H. A., and R. R. Sokal, 1973, *Numerical Taxonomy,* W. H. Freeman, San Francisco.

Snyder, J. P., and J. L. Gooch, 1973, Genetic differentiation in *Littorina saxatilis* (Gastropoda), *Mar. Biol.* (Berl.), 22 (2), 177–182.

Snyder, T. P., 1974, Lack of allozymic variability in three bee species, *Evolution,* **28,** 687–689.

Somero, G. N., and M. Soulé, 1974, Genetic variation in marine fishes as a test of the niche-variation hypothesis, *Nature,* **249,** 670–672.

Soulé, M., 1966, Trends in the insular radiation of a lizard, *Amer. Natur.,* **100,** 47–64.

Soulé, M., 1971, The variation problem: the gene flow-variation hypothesis, *Taxon,* **20** (1), 37–50.

Soulé, M., 1972, Phenetics of natural populations. III. Variations in insular populations of a lizard, *Amer. Natur.,* **106,** 429–446.

Soulé, M., 1973, The epistasis cycle: a theory of marginal populations, *Ann. Rev. Ecol. System.,* **4,** 165–187.

Soulé, M., S. Y. Yang, 1974, Genetic variation in side-blotcher lizards on islands in the Gulf of California, *Evolution,* **27,** 593–600.

Soulé, M., S. Y. Yang, M. G. Weiler, and G. C. Gorman, 1973, Island lizards: the genetic-phenetic variation correlation, *Nature,* **242,** 190–192.

Southern, E. M., 1975, Long range periodicities in mouse satellite DNA, *J. Mol. Biol.,* **94,** 51–69.

Sparrow, A. H., H. J. Price, and A. G. Underbrink, 1972, A survey of DNA content per cell and per chromosome of prokaryotic and eukaryotic organisms: some evolutionary considerations, in *Evolution of Genetic Systems,* H. H. Smith *et al.* (eds.), *Brookhaven Symp. Biol.,* **23,** 451–495.

Spassky, B., R. C. Richmond, S. Perez-Salas, O. Pavlovsky, C. A. Mourão, A. S. Hunter, H. Hoenigsberg, Th. Dobzhansky, and F. J. Ayala, 1971, Geography of the sibling species related to *Drosophila willistoni,* and the semispecies of the *Drosophila paulistorum* complex, *Evolution,* **25,** 129–143.

Speirs, J., and M. Birnstiel, 1974, Arrangement of the 5.8S RNA cistrons in the genome of *Xenopus laevis, J. Mol. Biol.,* **87,** 237–256.

Stanier, R. Y., D. Wachter, C. Gasser, and A. C. Wilson, 1970, Comparative immunological studies of two *Pseudomonas* enzymes, *J. Bacteriol.* **102,** 351–362.

Stanley, S. M., 1975, A theory of evolution above the species level, *Proc. Natl. Acad. Sci. U.S.A.,* **72,** 646–650.

Stebbins, G. L., 1950, *Variation and Evolution in Plants,* Columbia University Press, New York.

Stebbins, G. L., 1957, Self-fertilization and population variability in the higher plants, *Amer. Natur.,* **91,** 337–354.

Stebbins, G. L., 1966, Chromosomal variation and evolution, *Science,* **152,** 1463–1469.

Stebbins, G. L., 1971, *Chromosomal Evolution in Higher Plants,* Addison-Wesley, Reading, Massachusetts.

Stehli, F. G., R. G. Douglas, and N. D. Newell, 1969, Generation and maintenance of gradients in taxonomic diversity, *Science,* **164,** 947–949.

Stein, G. S., J. S. Stein, and L. J. Kleinsmith, 1975, Chromosomal proteins and gene regulation, *Sci. Amer.,* **232,** 46–57.

Strickberger, M. W., 1968, *Genetics,* Macmillan, New York.

Sturtevant, A. H., 1948, The evolution and function of genes, *Amer. Sci.,* **36,** 225–236.

Suomalainen, E., and A. Saura, 1973, Genetic polymorphism and evolution in parthenogenetic animals. I. Polyploid Curculionidae, *Genetics,* **74,** 489–508.

Sverdrup, H. L., M. W. Johnson, and R. H. Fleming, 1942, *The Oceans,* Prentice Hall, Englewood Cliffs, New Jersey.

Tappan, H., and A. R. Loeblich, Jr., 1973*a,* Evolution of the oceanic plankton, *Earth-Science Rev.,* **9,** 207–240.

Tappan, H., and A. R. Loeblich, Jr., 1973*b,* Smaller protistan evidence and explanation of the Permian-Triassic crisis, in *The Permian and Triassic Systems and Their Mutual Boundary,* A. Logan and L. V. Hills (eds.), Canadian Soc. Petrol. Geol., Mem. 2, pp. 465–480.

Tashian, R. E., R. J. Tanis, R. E. Ferrell, S. K. Stroup, and M. Goodman, 1972, Differential rates of evolution in carbonic anhydrase isozymes of catarrhine primates, *J. Hum. Evol.,* **1,** 545–552.

Tracey, M. L., and K. B. Nelson, 1975, Allozymic variation in the American lobster, *Homarus americanus, Genetics,* **80** (suppl.), 81.

Turner, B. J., 1974, Genetic divergence of Death Valley pupfish species: biochemical versus morphological evidence, *Evolution,* **28,** 281–294.

Tymowska, J., and M. Fischberg, 1973, Chromosome complements of the genus *Xenopus. Chromosoma,* **44,** 335–342.

Valentine, J. W., 1971, Resource supply and species diversity patterns, *Lethaia,* **4,** 51–61.

Valentine, J. W., 1972, Conceptual models of ecosystem evolution, in *Models in Paleobiology,* T. J. M. Schopf (ed.), Freeman, Cooper, San Francisco, pp. 192–215.

Valentine, J. W., 1973, *Evolutionary Paleoecology of the Marine Biosphere,* Prentice-Hall, Englewood Cliffs, New Jersey.

Valentine, J. W., 1975, Adaptive strategy and the origin of grades and groundplans, *Amer. Zool.,* **15,** 391–404.

Valentine, J. W., and F. J. Ayala, 1974, Genetic variation in *Frieleia halli,* a deep-sea brachiopod, *Deep-Sea Res.,* **22,** 37–44.

Valentine, J. W., and F. J. Ayala, 1976, Genetic variability in krill, *Proc. Nat. Acad. Sci. U.S.A.,* **73,** in press.

Van Valen, L., 1973, A new evolutionary law, *Evol. Theory,* **1,** 1–30.

Vasek, F., 1964, The evolution of *Clarkia unguiculata* derivatives adapted to relatively xeric environments, *Evolution,* **18,** 26–42.

Vasek, F., 1968, The relationship of two ecologically marginal, sympatric *Clarkia* populations, *Amer. Natur.,* **102,** 25–40.

Vaslet, C. A., and E. M. Berger, 1975, Intraspecific and interspecific ribosomal protein variation in *Drosophila, Genetics,* **80** (suppl.), 82.

Vermeij, G. J., 1973, Morphological patterns in high-intertidal gastropods: adaptive strategies and their limitations, *Mar. Biol.,* **20,** 319–346.

Vickery, R. K., 1974, Crossing barriers in the yellow monkey flowers of the genus *Mimulus* (Scrophulariaceae), in *Genetics Lectures,* Oregon State University Press, Corvallis.

Vigue, C. L., and F. M. Johnson, 1973, Isozyme variability in species of the genus *Drosophila.* VI. Frequency-property-environment relationships of allelic alcohol dehydrogenases in *D. melanogaster, Biochem. Genet.,* **9,** 213–227.

Vogt, V. M., 1973, Purification and further properties of single-strand-specific nuclease from *Aspergillus oryzae, Eur. J. Biochem.,* **33,** 192–200.

Wallace, B., 1968, *Topics in Population Genetics,* Norton, New York.

Wallace, B., and T. L. Kass, 1974, On the structure of gene control regions, *Genetics,* **77,** 541 558.

Wallace, D. C., and H. J. Morowitz, 1973, Genome size and evolution, *Chromosoma,* 40, 121–126.

Wallace, D. G., L. R. Maxson, and A. C. Wilson, 1971, Albumin evolution in frogs: a test of the evolutionary clock hypothesis, *Proc. Natl. Acad. Sci. U.S.A.,* **68,** 3127–3129.

Webb, D. S., 1969, Extinction-origination equilibria in late Cenozoic large mammals of North America, *Evolution,* **23,** 688–702.

Webster, T. P., R. K. Selander, and S. Y. Yang, 1972, Genetic variability and similarity in the *Anolis* lizards of Bimini, *Evolution,* **26,** 523–535.

Weisbrot, D. R., 1963, Studies on differences in the genetic architecture of related species of Drosophila, *Genetics,* **48,** 1121–1139.

Wetmur, J. G., and N. Davidson, 1968, Kinetics of renaturation of DNA, *J. Mol. Biol.,* **31,** 349–370.

Whitt, G. S., W. F. Childers, and P. L. Cho, 1973, Allelic expression at enzyme loci in an intertribal hybrid sunfish, *J. Hered.,* **64,** 55–61.

Wilson, A., 1975, Relative rates of evolution of organisms and genes, *Stadler Genet. Symp.,* **7,** 117–134.

Wilson, A. C., L. R. Maxson, and V. M. Sarich, 1974*a,* Two types of molecular evolution. Evidence from studies of interspecific hybridization, *Proc. Natl. Acad. Sci. U.S.A.* **71,** 2843–2847.

Wilson, A. C., and E. M. Prager, 1974, Antigenic comparison of lysozymes, in *Lysozyme,* E. F. Osserman, R. E. Canfield, and S. Beychock (eds.), Academic Press, New York, pp. 127–141.

Wilson, A. C., and V. M. Sarich, 1969, A molecular time scale for human evolution, *Proc. Natl. Acad. Sci. U.S.A.,* **63,** 1088–1093.

Wilson, A. C., V. M. Sarich, and L. R. Maxson, 1974*b,* The importance of gene rearrangement in evolution: evidence from studies on rates of chromosomal, protein, and anatomical evolution, *Proc. Natl. Acad. Sci. U.S.A.,* **71,** 3028–3030.

Wolff, S., S. Abrahamson, M. A. Bender, and A. D. Conger, 1974, The uniformity of normalized radiation-induced mutation rates among different species, *Genetics,* **78,** 133–134.

Wright, S., 1932, The roles of mutation, inbreeding, crossbreeding and selection in evolution, *Proc. VI Intern. Cong. Genet.,* **1,** 356–366.

Yamazaki, T., 1971, Measurement of fitness at the esterase-5 locus in *Drosophila pseudoobscura, Genetics,* **67,** 579–603.

Yang, S. Y., M. Soulé, and G. C. Gorman, 1974, *Anolis* lizards of the Eastern Caribbean: a case study in evolution. I. Genetic relationships, phylogeny and colonization sequence of the *roquet* group, *System. Zool.,* **23,** 387–399.

Zouros, E., 1973, Genetic differentiation associated with the early stages of speciation in the *Mulleri* subgroup of *Drosophila, Evolution,* **27,** 601–621.

Zouros, E., 1975, Electrophoretic variation in allozymes related to function or structure?, *Nature,* **254,** 446–448.

Zouros, E., and C. B. Krimbas, 1969, The genetics of *Dacus oleae.* III. Amount of variation at two esterase loci in a Greek population, *Genet. Res.,* **14,** 249–258.

Zouros, E., C. B. Krimbas, S. Tsakas, and M. Laukas, 1974, Genic *versus* chromosomal variation in natural populations of *Drosophila subobscura, Genetics,* **78,** 1223–1244.

Zuckerkandl, E., 1965, The evolution of hemoglobin, *Sci. Amer.,* **212**(5), 110–118.

Zuckerkandl, E., and L. Pauling, 1965, Evolutionary divergence and convergence in proteins, in *Evolving Genes and Proteins,* V. Bryson and H. J. Vogel (eds.), Academic Press, New York, pp. 97–166.

AUTHOR INDEX

SUBJECT INDEX

ABOUT THE BOOK

This book was set in Linotype Baskerville at V & M Typographical. The editor was Robert H. Warner, Jr. The designer was Joseph Vesely. Illustrations were drawn at Vantage Art, Inc. Complete manufacturing was done by The Murray Printing Company.